Roadmap
to 8th Grade
Math:
NEW YORK EDITION

Roadmap
to 8th Grade
Math:
NEW YORK EDITION

by
Karen Lurie

Random House, Inc.
New York

This workbook was written by The Princeton Review, one of the nation's leaders in test preparation. The Princeton Review helps millions of students every year prepare for standardized assessments of all kinds. The Princeton Review offers the best way to help students excel on standardized tests.

The Princeton Review is not affiliated with Princeton University or Educational Testing Service.

Princeton Review Publishing, L.L.C.

2315 Broadway

New York, NY 10024

E-mail: textbook@review.com

Copyright 2003 by Princeton Review Publishing, L.L.C.

ISBN 0-375-76355-4

Editor: Russell Kahn

Designer: Greta Englert

Production Editor: Lisbeth Dyer

Development Editor: Scott Bridi

Manufactured in the United States of America on partially recycled paper.

9 8 7 6 5 4 3 2 1

CONTENTS

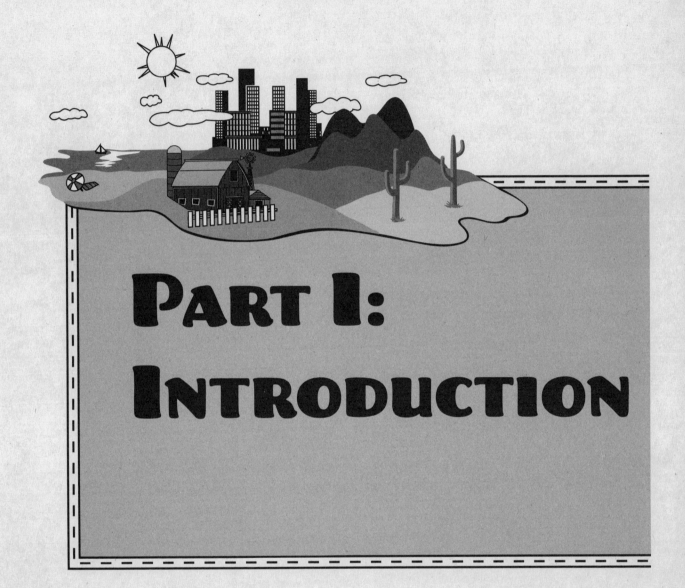

PART I:

INTRODUCTION

INTRODUCTION FOR PARENTS AND TEACHERS

About The Princeton Review

The Princeton Review is one of the nation's leaders in test preparation. We prepare more than 2 million students every year with our courses, books, on-line services, and software programs. In addition to helping New York students with their Grade 4 and Grade 8 tests, we make study guides for the high school Regents series. We also coach students around the country on many other statewide standardized tests as well as on college entrance exams, such as the SAT-I, SAT-II, PSAT, and ACT. Our strategies and techniques are unique and, most importantly, successful. Our goal is to reinforce skills that students have been taught in the classroom and show them how to apply those skills to the specific format and structure of the New York State test.

About This Book

Roadmap to 8th Grade Math: New York Edition contains three main parts: an introduction, lessons in which math skills are reviewed, and practice tests.

The introduction explains the particulars of the New York State Grade 8 Math test, including vital information such as the tools students are allowed to use while taking the test and the amount of time they will have to finish the test.

Each lesson (or "mile") focuses on a specific set of skills that is included in the New York State eighth-grade math curriculum. The twelve miles in this book include the skills that will be tested on the New York State Grade 8 Math test. The first two miles provide techniques for answering the different types of questions that will appear on the test: multiple choice, short response, and extended response.

This book also contains two full-length practice tests. Reviewing students' performance on the tests will help you assess which skills they need more practice with before they take the New York State Grade 8 Math test. An introduction to the practice tests precedes Practice Test One. (Answer keys and explanations for the two practice tests begin on page 117 for Practice Test One and page 187 for Practice Test Two.) By administering the practice tests under test-like conditions, you will help students become familiar with the testing situation they will experience when they take the New York State Grade 8 Math test.

Before students begin working through this book, take a moment to review the table of contents and to look through the content of the lessons. We realize that every student, and every class, has different strengths and weaknesses. There will be no harm done if students work on the miles out of sequence.

About the New York State Testing Program

All students in New York State must take tests in mathematics, English language arts (ELA), and science in grades 4 and 8. Grade 5 and grade 8 students also must take a test in social science. Teachers and school administrators use the test scores as a factor in determining whether a student should be promoted to the next grade. Test scores can also be a factor in deciding if a student needs extra instruction during school, after school, or in summer school. In fact, students who score below Level 3 on the Grade 8 Math or ELA tests will receive "instructional intervention," according to the New York State Education Department. The type of intervention varies from one school district to another.

About the New York State Grade 8 Math Test

The New York State Grade 8 Math test is administered over a two-day period in early May. Check your school's schedule for the exact dates.

In Session 1 of the Grade 8 Math test, there are two parts. The first part includes 27 multiple-choice questions. All multiple-choice questions must be answered on a separate answer sheet. (See page 78.) The second part includes about 6 open-response questions, which will be answered directly in the test book. Students will have 35 minutes to complete each part of Session 1, and they will have a short break between the two parts of the session. Students will have 70 minutes to answer all of the questions in Session 1.

In Session 2 of the Grade 8 Math test, there are about 12 open-response questions. Open-response questions include two- and three-point open-response questions. Students have 70 minutes to answer all of the questions in Session 2. A short break midway through Session 2 will not count against the time allotted for the session. Responses to the questions in Session 2 should be written directly in the test book.

For both sessions of the test, students are allowed to write in their test books. This enables students to write their calculations next to the problems they are solving.

Students will be given a ruler and a protractor to use while they take the test. We've provided similar tools for students on page 79. Students should cut out these tools and use them as they work on the practice tests in this book.

Students will be allowed to use a calculator and a reference sheet when they answer the open-response questions on the test. We recommend a scientific calculator for the test. We've provided a reference sheet (on page 81) similar to the one that students will see on the New York Grade 8 Math test.

Students with disabilities may be allowed certain special accommodations. Information concerning students with disabilities may be found on the Internet at ftp://unix2.nysed.gov/pub/education.dept.pubs/vesid/oses/test.access.mod/ testacce.txt. For more information, visit www.emsc.nysed.gov.

STUDENT INTRODUCTION

About This Book

This book is designed to help you review what you need to know for your eighth-grade math class. It will also help you improve your score on the Grade 8 Math test that you will take in the spring.

There are three main parts of this book: the introduction (which you are reading now), the miles (which review math skills), and the practice tests (which are similar to the New York State Grade 8 Math test).

About the New York State Grade 8 Math Test

All eighth-grade students in New York State have to take a Grade 8 Math test. The purpose is to show your teachers and parents what you know. The test results will tell them which math skills you know well, and which ones you might need a little help with.

You will take the test during a two-day period sometime in May while you are in the eighth grade. (You can ask your teacher for the exact testing dates if you don't know them.) Each of the two days includes 70 minutes of testing. During that time, you will have to answer about 45 questions in total. Of those questions, about 27 will be multiple choice and about 18 will be open response (short response and extended response).

About the Multiple-Choice Questions

Multiple-choice questions have four answer choices (A, B, C, and D or F, G, H, and J). To answer these questions, pick the best answer choice, and mark it with your pencil on the separate bubble sheet. Make sure to answer every question because you can't earn credit for a blank answer. See Mile 1 (on page 7) for more information about answering these types of questions.

About the Open-Response Questions

Open-response questions require you to write out your calculations and the final answer. For the open-response questions, you can earn partial credit even if your final answer is incorrect. That's why it's very important to show your work. Write clearly; if the grader can't read your handwriting, you may lose out on points. See Mile 2 (on page 9) for more information about answering these types of questions.

Tools

You will be given a protractor and a ruler to help you answer the questions on the test. You will also be allowed to use a reference sheet with formulas and a four-function or scientific calculator for the open-response questions on the test.

Preparing for the New York State Grade 8 Math Test

Here is a list of things you can do to prepare for the New York State Grade 8 Math test.

- **Ask questions.** If you are confused after you finish working on a mile (or even answering just one question), ask a parent or teacher for help. Asking questions is the best way to make sure that you understand what you have to do in order to do well on the test.

- **Read.** Read everything you can. Read newspapers, magazines, books, plays, poems, comics, and even the back of your cereal box. The more you read, the better you will become at it. And the better you read, the more likely you will do well on the Grade 8 Math test. Many of the questions on the math test will be word problems. If you can't read and understand the problems, you won't have a chance to show all the math you know.

- **Practice.** Math is all around you. Every time you go to the grocery store, make dinner, or play softball or baseball, math comes along for the ride. Use real-life opportunities to practice the math you know. Add up the prices of the groceries in your cart while you're waiting on line. See if your addition matches what comes up on the register. If you play baseball on a team, keep track of your batting average. Every day there will be more math to do, and the great thing is that it's the same type of math that will be on your Grade 8 Math test.

- **Eat well and get a good night's sleep.** Your body doesn't work well when you don't eat good food and get enough sleep. Neither does your brain. On the night before the test, make sure to go to bed at your normal time and get plenty of sleep. You should also eat a healthy breakfast on the morning of the testing day. It is important to be awake and alert while you take the Grade 8 Math test, or any other test for that matter.

This is just the beginning of the road. There are great things to learn ahead. So buckle your seatbelt and get ready to travel the first mile to New York math excellence.

MILE 1: ANSWERING MULTIPLE-CHOICE QUESTIONS

The New York State Grade 8 Math test consists of 45 questions, 27 of which are multiple choice. Each multiple-choice question is followed by four possible answer choices. The choices could be A, B, C, and D or F, G, H, and J. You have to pick the best answer choice for each question.

You should approach multiple-choice questions like any other math problems. Carefully read the question and figure out what you need to do in order to answer it. You may have to perform calculations or use reasoning to find the answer. What makes multiple-choice questions different is that they come with four answer choices; one of them is the correct answer. Once you figure out the answer, look for it among the answer choices. When you find it, just fill in the bubble that goes with it.

But what happens if it's not as easy as that? What if you have trouble figuring out the answer to a question? That's where the Process of Elimination comes in.

The Process of Elimination (POE)

It's great that so many questions on the Grade 8 Math test are multiple choice. Why? Because even if you're not sure which answer choice is correct for a multiple-choice question, you can often use the Process of Elimination (POE) to find it. How does this work? If you're having difficulty finding the correct answer choice, look for answer choices that you *know* are wrong. Then get rid of those choices by crossing them off with your pencil. Look at an example to see how this works.

▶ What is the capital of Burundi?

_____

Not sure? Without any answer choices, you would have to guess (and you would probably be wrong). But with multiple-choice questions like the ones on the Grade 8 Math test, you just have to pick the answer choice that you think is best. The answer choices often give valuable information. Now look at the answer choices.

▶ What is the capital of Burundi?

A   New York City
B   Bujumbura
C   London
D   Albany

You probably know that New York City isn't the capital of Burundi. That means it can't be the right answer choice. Cross off answer choice (A). London is the capital of England, so choice (C) can't be right either. Albany, choice (D), is the capital of New York State. Eliminate it. Even if you've never even *heard* of Burundi, you can probably recognize that (A), (C), and (D) are *not* its capital. Singling out Bujumbura as the correct answer choice is an example of using POE.

POE works especially well with math questions. Look at a question that could be difficult to solve without using POE.

▶ Adam and Carmen made $1,221.58 by selling toys in their toy store over a one-week period. They made a promise to give 10% of the money they made to charity. After they give money to charity, about how much will Adam and Carmen have left from the one-week period?

A $950
B $1,100
C $1,220
D $1,350

If you had a hard time figuring out the right answer choice, it helps to attack the answer choices. Adam and Carmen gave 10% of the money they made. That would mean that they would have less than what they started with. The value in answer choice (D) is too big to be correct. Likewise, answer choice (C) is about the amount of money that Adam and Carmen made in the first place. If they gave 10% of the money away, they must have less than that! Answer choice (C) is wrong, too. Cross it off.

Now you're down to only two answer choices. At this point, choices (A) and (B) are both reasonable; they both list smaller values than what Adam and Carmen started off with. However, 10% of $1,221.58 is about $122; $1,221.58 – $122 is about $1,100. The correct answer choice is (B). Adam and Carmen were left with about $1,100 after giving 10% of the money they made to charity.

What if you can't get rid of all the wrong answer choices? That's okay. You will have a better chance of picking the correct answer choice if you can get rid of even one wrong answer choice. You always start with four answer choices. If you have no idea how to answer a question, you have only a 25% chance of guessing the correct one. If you get rid of one or two answer choices, your odds of picking the correct answer improve to either 33% or 50%.

Although you probably won't be able to eliminate all three wrong answers for *every* question, crossing out even one or two incorrect answer choices can improve your score dramatically on any multiple-choice test that you take.

POE can have a significant impact on your test score. You should combine your mathematical skills with POE to maximize your score on every test.

MILE 2: ANSWERING OPEN-RESPONSE QUESTIONS

The New York State Grade 8 Math test consists of 45 questions, about 18 of which are open response. Each open-response question requires you to write your answer on a line or a few lines in your test book. Some open-response questions, called short-response questions, will be worth two points. Other open-response questions, called extended-response questions, will be worth three points. The different types of open-response questions look the same, but extended-response questions usually have an extra step to solve.

Short-response and extended-response questions require you to:

- Solve problems

- Make comparisons, interpretations, and predictions

- Discuss concepts

- Demonstrate problem-solving strategies

While there are no answer choices to help you find the answer to open-response questions, there are a few other useful things:

- A calculator

- The mathematics reference sheet

- The opportunity to earn partial credit

Show Your Work!

The most important thing to remember when answering open-response questions is to always show your work. Even if you got the answer correct, you will not receive full credit if you don't show how you found it. And even if you didn't give the right answer, you might get partial credit for showing your work.

The more work you show, the better off you will be. When solving open-ended questions, it may help to remember this little phrase: When in doubt, write it out. Because you can use a calculator on the open-ended questions, the focus is not on the calculations, but on the process of—and logic behind—solving the problems.

Scoring

Open-response questions will be scored using details of how you solved a problem as well as your final solution. The key is whether you are able to show that you understand how to solve the problem. The graders want to see your work, and they want to understand what you did to solve the problem. Look at an example on the next page.

▶ Jill went to the candy store to buy some chocolates for her birthday party. The candy store sells boxes of 6 pieces of chocolate for $5.00. Jill expects to have 8 people at her party (including herself) and spends $20.00 on the boxes of chocolates. If she wants each person to have the same number of chocolates, how many pieces of chocolate can each person at the party have?

Show your work.

Answer _____ pieces of chocolate

There are many ways to correctly solve this problem. Below is a series of steps that would lead to the correct answer and would be awarded full credit.

One box of chocolates = $5.00. Jill spends $20.00 on boxes of chocolates.

$20.00 ÷ $5.00 = 4.

Jill bought 4 boxes of chocolates.

Each box of chocolates contains 6 chocolates.

4 boxes of chocolates × 6 = 24 chocolates.

Jill bought 24 pieces of chocolate.

There will be 8 people at Jill's party (including Jill).

24 pieces of chocolate ÷ 8 people at the party = 3.

Each person at the party can have 3 pieces of chocolate.

Answer _____24_____ pieces of chocolate

That's it! Because this solution showed all the calculations involved in finding the right answer, it would be awarded full credit. (This short-answer question would be worth two points.)

Always read every question carefully. If you misread the question, you may do all the right calculations but still write the wrong answer. Give yourself enough time to identify what the question is really asking.

PART II:
MATH REVIEW

It's time to start the review of grade 8 math. Even if the material is familiar to you, read through it anyway. The better you know this stuff, the more likely you'll nail the questions on the New York State Grade 8 Math test.

INTEGERS

Integers are whole numbers that you might see on a number line. Examples of integers are –5, –4, –3, –2, –1, 0, 1, 2, 3, 4, and 5.

Fractions and decimals such as $\frac{1}{2}$ or 4.3 are not integers.

Positive integers are integers that are larger than zero.

Negative integers are integers that are smaller than zero.

Zero itself is neither positive nor negative. Zero has some interesting properties.

- The sum (addition) of zero and any number is that number itself.
- The product (multiplication) of zero and any number is zero.
- Zero is even.

Even integers are integers that can be divided by 2 with no remainder, such as –4, –2, 0, 2, 4, and 6.

Odd integers are integers that cannot be divided by 2 with no remainder, such as –3, –1, 1, 3, and 5.

Adding and Subtracting Integers

A **sum** is the result of addition, and a **difference** is the result of subtraction.

The **Associative Law of Addition** states that when you are adding a series of numbers, you can regroup the numbers in any way you'd like. In other words:

$$a + (b + c) = (a + b) + c = (a + c) + b$$
$$4 + (5 + 8) = (4 + 5) + 8 = (4 + 8) + 5$$

The **Commutative Property of Addition** states that when you are adding real numbers, order doesn't matter. In other words:

$$a + b = b + a$$
$$4 + 5 = 5 + 4$$

Multiplying and Dividing Integers

A **product** is the result of multiplication, and a **quotient** is the result of division. The product of two integers with the same sign is positive. The product of two integers with different signs is negative.

positive $\times$ positive = positive	For example, $2 \times 2 = 4$
negative $\times$ negative = positive	For example, $-2 \times -2 = 4$
positive $\times$ negative = negative	For example, $2 \times -2 = -4$

The quotient of two integers with the same sign is positive. The quotient of two integers with different signs is negative.

positive $\div$ positive = positive	For example, $4 \div 2 = 2$
negative $\div$ negative = positive	For example, $-4 \div -2 = 2$
positive $\div$ negative = negative	For example, $4 \div -2 = -2$
negative $\div$ positive = negative	For example, $-4 \div 2 = -2$

The **Associative Law of Multiplication** states that when you are multiplying a series of numbers, you can regroup the numbers in any way you'd like (just like addition). In other words:

$$a \times (b \times c) = (a \times b) \times c = (a \times c) \times b$$
$$4 \times (5 \times 8) = (4 \times 5) \times 8 = (4 \times 8) \times 5$$

The **Commutative Property of Multiplication** states that when you are multiplying real numbers, order doesn't matter. In other words,

$$a \times b = b \times a$$
$$4 \times 5 = 5 \times 4$$

Multiplying and Dividing Decimals

To multiply two numbers if one or both of them has a decimal, multiply as if they were integers.

$$
\begin{array}{r}
1.25 \\
\times\ 1.5 \\
\hline
1875
\end{array}
$$

Count the decimal places to the right of the decimal in the problem. Then put that many digits to the right of the decimal in your answer. In this case, there are three (two from 1.25 and one from 1.5). Put a decimal point three units to the left: 1.875

To divide two numbers if one or both is a decimal, the divisor needs to be a whole number. You can change the divisor into a whole number by multiplying both divisor and dividend by a power of 10 (10; 100; 1,000; etc.).

$$22.05 \div 4.2 = 42\overline{)220.5}^{\,5.25}$$

ORDER OF OPERATIONS

The **order of operations** is the order in which operations are to be performed. It is: **P**arentheses, **E**xponents, **M**ultiplication and **D**ivision, and **A**ddition and **S**ubtraction. First you perform whatever operations are inside the parentheses; then you take care of the exponents; then you perform all multiplication and division at the same time, from *left* to *right*, followed by addition and subtraction, from *left* to *right*. You can remember this order as **P**lease **E**xcuse **M**y **D**ear **A**unt **S**ally. That's **PEMDAS,** for short.

Here's an example.

$$10 - (6 \div 6) - (3 + 3) - 3 =$$

Start with the parentheses. The expression inside the first pair of parentheses, $6 \div 6$, equals 1. The expression inside the second pair equals 6. Now rewrite the problem as follows:

$$10 - 1 - 6 - 3 =$$
$$9 - 6 - 3 =$$
$$3 - 3 =$$
$$= 0$$

FACTORS

A number x is a **factor** of y if y can be divided by x without leaving a remainder. For example, 1, 2, 3, 4, 6, and 12 are all factors of 12. If you always start with 1 and the number itself when you write down the factor pairs, you won't forget any of them.

1 and 12

2 and 6

3 and 4

DIVISIBILITY

If a number is a factor of a given number, the given number is **divisible** by the factor. Some divisibility shortcuts to remember are on the next page.

- An integer is divisible by 2 if its units digit is divisible by 2.
- An integer is divisible by 3 if the sum of its digits is divisible by 3.
- An integer is divisible by 4 if its last two digits form a number divisible by 4.
- An integer is divisible by 5 if its units digit is either 0 or 5.
- An integer is divisible by 6 if it is divisible by *both* 2 *and* 3.
- An integer is divisible by 9 if the sum of its digits is divisible by 9.
- An integer is divisible by 10 if its units digit is 0.

PRIME FACTORIZATION

A whole number greater than 1 with exactly two factors, itself and 1, is a **prime number.** A whole number greater than 1 with more than 2 factors is a **composite number.** The numbers 0 and 1 are neither prime nor composite: 0 has infinite factors, and 1 only has one factor, itself. The number 2 is the only even prime number. A number expressed as a product of factors that are all prime is called the **prime factorization** of the number. For example, the prime factorization of 12 is $2 \times 2 \times 3$.

Try the following example:

▶ Find the prime factorization of 306.

Use divisibility rules to help you. Note that the digits in 306 add up to 9. That means that 306 is divisible by 9. $306 = 34 \times 9$. 9 can be broken down into 3×3. So you have $34 \times 3 \times 3$. 34 is divisible by 2. That makes $17 \times 2 \times 3 \times 3$. 17 is a prime number, so that's it.

ABSOLUTE VALUE

The **absolute value** of a number is the positive value of that number.

The absolute value of 5 (written as $|5|$) is 5.

The absolute value of –5 (written as $|-5|$) is 5.

Exercise I

Practice by solving the following problems (answers on page 75):

a. $|-8| + 3$

b. $3 + 4 \times 5 - 7$

c. $2 \times (5 \times 9) + 6$

d. $-5 \times (-2 \times 3)$

e. Which of the following numbers is divisible by 3?
   893, 406, 624, or 521?

f. What are the factors of 64?

g. What are the prime factors of 48?

Mile 4: Factors, Multiples, and Fractions

GREATEST COMMON FACTOR

The greatest of the factors common to two or more numbers is called the **greatest common factor (GCF)** of the numbers.

Here's an example.

▶ Find the greatest common factor of 24 and 40.

The factors of 24 are 1, 2, 3, 4, 6, 8, 12, and 24.

The factors of 40 are 1, 2, 4, 5, 8, 10, 20, and 40.

The greatest factor common to both is 8.

Here's another example.

▶ Use prime factorization to find the greatest common factor of 90, 54, and 81.

The factors of 90 are 1, 2, 3, 5, 6, 9, 10, 15, 18, 30, 45, and 90.

The factors of 54 are 1, 2, 3, 6, 9, and 54.

The factors of 81 are 1, 3, 9, 27, and 81.

The greatest factor common of 54, 81, and 90 is 9.

LEAST COMMON MULTIPLE

A **multiple** of a number is the product of that number and any whole number. The least of the nonzero common multiples of two or more numbers is called the **least common multiple (LCM).**

Here's an example.

▶ Find the least common multiple of 4, 6, and 8.

Start with the multiples of 4: 4, 8, 12, 16, 20, 24, 28, 32, and so on.
Then find the multiples of 6: 6, 12, 18, 24, 30, 36, 42, 48, 54, 60, and so on.
So far, the least common multiple is 12 because 12 is the smallest number that matches in both. Finally, find the multiples of 8: 8, 16, 24, 32, 40, and so on.
There is no 12, but in the multiples of 4, 6, and 8, the least common multiple is 24 because 24 is the smallest number that matches in all three.

RATIONAL NUMBERS

A **rational number** is any number (positive or negative) that can be expressed in the form $\frac{a}{b}$, where a and b are integers and $b \neq 0$. $\frac{3}{7}$, 1.125, and 2,146 are examples of rational numbers. When a rational number is expressed as a fraction, it is often expressed in its simplest, or reduced, form; a fraction is in its simplest form if the greatest common factor of the numerator and denominator is 1.

Here is an example.

▶ Write $\frac{30}{45}$ in simplest form.

The greatest common factor of 30 and 45 is 15. Dividing $\frac{30}{45}$ by $\frac{15}{15}$ reduces the fraction to $\frac{2}{3}$.

A **terminating decimal** is a decimal whose digits end, such as 0.45. Every terminating decimal can be written as a fraction with a denominator of 10, 100, 1,000, and so on. A **repeating decimal** is a decimal whose digits repeat in groups of one or more, such as 0.181818 . . . . They can be expressed with **bar notation,** when a bar is placed over the digits that repeat:

$$0.\overline{18}$$

FRACTIONS

Adding and Subtracting Like Fractions

To add fractions with like denominators, just add the numerators. To subtract fractions with like denominators, just subtract the numerators. For example, $\frac{8}{15} - \frac{4}{15} = \frac{4}{15}$.

A **mixed number** is a number that is represented as an integer and a fraction, like this: $2\frac{3}{4}$. To change a mixed number into an **improper fraction,** multiply the denominator of the fraction by the number, add that sum to the numerator, and put the whole thing back over the denominator. In the case of $2\frac{3}{4}$, 4 (the denominator) × 2 (the number) = 8. 8 + 3 (the old numerator) = 11 (the new numerator). The result, $\frac{11}{4}$, has the same value as $2\frac{3}{4}$. The difference is that $\frac{11}{4}$ is easier to work with.

Adding and Subtracting Unlike Fractions

To find the sum or difference of two fractions with unlike denominators, rename the fractions with a common denominator. Then add or subtract and simplify. Here's an example.

▶ Evaluate $m - n$ if $m = \dfrac{4}{5}$ and $n = \dfrac{3}{4}$

This question asks you to subtract $\dfrac{3}{4}$ from $\dfrac{4}{5}$. To add or subtract fractions, you need to find a common denominator. The easiest way to do this is to use the bowtie:

$$\dfrac{4}{5} \bowtie \dfrac{3}{4} \qquad \dfrac{16}{20} - \dfrac{15}{20}$$

To use the bowtie, you multiply the two denominators (numbers on the bottom). Multiply up and diagonally to get the numerators. Then you will have two fractions with a common denominator, and can add or subtract normally.

If you change $\dfrac{3}{4}$ into a fraction with a denominator of 20, you need a new numerator that is the result of multiplying the numerator by the same number that we had to multiply the denominator by to get 20. $4 \times 5 = 20$, so multiply 3 by 5 and get 15. $\dfrac{3}{4}$ becomes $\dfrac{15}{20}$.

Now, the question is $\dfrac{16}{20} + \dfrac{15}{20}$. Now that the denominators are the same, just add the numerators and get $\dfrac{31}{20}$.

Multiplying Fractions

To multiply fractions, multiply the numerators straight across, multiply the denominators, and simplify. If possible, reduce any fractions before you multiply.

$$\dfrac{4}{5} \times \dfrac{2}{3} = \dfrac{8}{15}$$

Dividing Fractions

To divide fractions, just change the product of a rational number and its **multiplicative inverse** (also known as the reciprocal) is 1. For example, the multiplicative inverse of 2 is $\frac{1}{2}$, and $2 \times \frac{1}{2} = 1$.

Try this example.

▶ Name the multiplicative inverse of $3\frac{1}{4}$.

First, convert $3\frac{1}{4}$ to an improper fraction: $\frac{13}{4}$. Now flip it over (that is, put its denominator over its numerator), and you get $\frac{4}{13}$.

To divide by a fraction, multiply by its multiplicative inverse (or reciprocal). In other words, flip the second fraction over, then multiply.

$$\frac{4}{5} \div \frac{2}{3} = \frac{4}{5} \times \frac{3}{2} = \frac{12}{10} = \frac{6}{5}$$

Comparing and Ordering Fractions

Sometimes you'll be asked to find the greatest fraction or put fractions in order from least to greatest.

Here's an example.

▶ Put the following fractions in order from least to greatest:

$$\frac{2}{3}, \frac{3}{4}, \frac{1}{3}, \frac{2}{4}$$

The easy way to do this is to give all of the fractions a common denominator. Make 12 the new denominator of the four fractions and compare.

$$\frac{8}{12}, \frac{9}{12}, \frac{4}{12}, \frac{6}{12}$$

Now that the denominators are all the same, all you have to do is put the numerators in order from least to greatest.

$$\frac{4}{12}, \frac{6}{12}, \frac{8}{12}, \frac{9}{12}$$

This means that the order of the original fractions from least to greatest would be.

$$\frac{1}{3}, \frac{2}{4}, \frac{2}{3}, \frac{3}{4}$$

VARIABLES, EXPRESSIONS, AND EQUATIONS

First, review a few basics of math language and symbols.

$3(2) = 3 \times 2$

$5x = 5 \times x$

$xy = x \times y$

$4 \times 7a = 4 \times 7 \times a$

$8ab^2 = 8 \times a \times b \times b$

$a[b(cd)] = a \times [b \times (c \times d)]$

$\dfrac{x}{10z} = x \div (10 \times z)$

$a\left(\dfrac{b}{3}\right) = a \times (b \div 3)$

Now review writing expressions and equations.

five less than a number	$x - 5$
a number increased by 12	$x + 12$
twice a number decreased by 3	$2x - 3$
the quotient of a number and 4	$\dfrac{x}{4}$

- Problems involving addition will mention things like *plus, sum, more than, increased by, total,* or *in all.*

- Problems involving subtraction will mention things like *minus, difference, less than, subtract,* or *decreased by.*

- Problems involving multiplication will mention things like *times, product, multiplied, each,* or *of.*

- Problems involving division will mention things like *divided, quotient,* or *separated.*

When you're doing word problems, sometimes your task is to form equations out of the words. For example, "24 is 6 more than a number" can be rewritten as $24 = x + 6$. "Five times a number is 60" can be rewritten as $5x = 60$.

DISTRIBUTION AND FACTORING

The **Distributive Law** states that the sum of two addends (any number to be added) multiplied by a number is the sum of the product of each addend and the number. In other words,

$$a(b + c) = ab + ac$$
$$a(b - c) = ab - ac$$

Try this example.

▶ What is the value of x if 12(66) + 12(24) = x?

Using the Distributive Law, x must equal 12(66 + 24), or 12(90) = 1,080.

When you use the Distributive Law to rewrite the expression $xy + xz$ in the form $x(y + z)$, you are *factoring* the original expression. That is, you take the factor common to both terms of the original expression (x), and "pull it out." This gives you a new, "factored" version of the expression you began with.

When you use the Distributive Law to rewrite the expression $x(y + z)$ in the form $xy + xz$, you are *unfactoring* the original expression.

SOLVING EQUATIONS

Any equation with one variable can be solved by manipulating the equation. You get the variables on one side of the equation, and the numbers on the other side. To do this, you can add, subtract, multiply, or divide both sides of the equation by the same number. Just remember that anything you do to one side of an equation, you *have* to do to the other side. Be sure to write down every step.

Look at an example.

▶ $4x - 3 = 9$

You can get rid of negatives by adding something to both sides of the equation, just as you can get rid of positives by subtracting something from both sides of the equation.

$$
\begin{array}{rl}
4x - 3 & = 9 \\
+ 3 & + 3 \\
\hline
4x & = 12
\end{array}
$$

You may already see that $x = 3$. But don't forget to write down that last step. Divide both sides of the equation by 4.

$$\frac{4x}{4} = \frac{12}{4}$$

$$x = 3$$

Exercise 2

Try the following problems (answers on page 75):

a. What is the greatest common factor of 16 and 24?

b. What is the least common multiple of 4 and 6?

c. What is $\dfrac{5}{8} + \dfrac{6}{16}$?

d. What is $\dfrac{3}{4} - \dfrac{1}{3}$?

e. What is $\dfrac{2}{3} \times \dfrac{3}{8}$?

f. What is $\dfrac{1}{2} \div \dfrac{1}{4}$?

g. If 8 is 2 more than $2x$, what is the value of x?

h. If $3x + 5 = x + 3$, what is the value of x?

MILE 6: FRACTIONS, DECIMALS, AND PERCENTS

Memorize the percentage-decimal-fraction equivalents below; you can use these friendly fractions and percentages to help you estimate answers.

DECIMAL	FRACTION	PERCENT
0.01	$\frac{1}{100}$	1%
0.1	$\frac{1}{10}$	10%
0.2	$\frac{1}{5}$	20%
0.25	$\frac{1}{4}$	25%
$0.\overline{33}$	$\frac{1}{3}$	$33\frac{1}{3}\%$
0.4	$\frac{2}{5}$	40%
0.5	$\frac{1}{2}$	50%
0.6	$\frac{3}{5}$	60%
$0.\overline{66}$	$\frac{2}{3}$	$66\frac{2}{3}\%$
0.75	$\frac{3}{4}$	75%
0.8	$\frac{4}{5}$	80%
1.0	$\frac{1}{1}$	100%
2.0	$\frac{2}{1}$	200%

To convert a decimal to a percent, just move the decimal point two places to the right and add the % sign. This turns 0.8 into 80%, 0.25 into 25%, 0.5 into 50%, and 1 into 100%. Try this example.

► Express 70% as a fraction and a decimal.

70% is the same as $\frac{70}{100}$, which can be reduced to $\frac{7}{10}$. 70% is also the same as 0.70. Here's another one.

▶ Express $\frac{4}{5}$ as a percentage.

Because $\frac{4}{5}$ is the same as 4 ÷ 5, calculate 4 ÷ 5.

$$5\overline{)4.0}^{\,0.8}$$

You can also set $\frac{4}{5}$ equal to $\frac{x}{100}$, like this.

$$\frac{4}{5} = \frac{x}{100}$$

Then multiply both sides of the equation by 100 to get the percent.

$$\frac{400}{5} = x$$

$$80 = x$$

So $\frac{4}{5}$ is the same as 80%.

PERCENT EQUATIONS

Here's an equation you can use when dealing with percents: percentage (P) = rate (R) times the base (B), or $P = RB$. The rate can be expressed as a fraction or a decimal. The base is the original number.

Here's an example.

▶ Find 30% of 540.

$$P = RB$$
$$P = (0.30)(540)$$
$$P = 162$$

Here's another one.

▶ 15 is 60% of what number?

$$P = RB$$

$$15 = (0.60)(B)$$

$$\frac{15}{0.60} = B$$

$$25 = B$$

Percents can also be used to express a fractional part of something. For this, you can use the percent proportion formula: $\frac{\text{percentage}}{\text{base}} = \frac{\text{rate}}{100}$ or $\frac{P}{B} \times \frac{R}{100}$.
Here's an example.

▶ Ling orders a pizza with 8 equal slices, and she eats
three of them. What percent of the pizza has she eaten?

Use $\frac{P}{B} = \frac{R}{100}$, where P is the number of slices Ling has eaten, B is the total number of slices, and R is the value that you are looking for.

$$\frac{3}{8} = \frac{R}{100}$$

Now cross multiply.

$$8R = (3)(100)$$

$$8R = 300$$

$$R = \frac{300}{8}$$

$$R = 37.5$$

Ling has eaten 37.5% of the pizza.

MORE ON PERCENTS

To quickly find 1% of a number, move the decimal point to the left 2 places.

To quickly find 10% of a number, move the decimal point to the left 1 place. This can also help you estimate to find a correct answer choice.

If you have trouble setting up percent equations with the $P = RB$ formula, try using **translation.** Translating a problem lets you express it as an equation, which is easier to manipulate. Here's a word problem dictionary that will help with your translations.

If you see the word...	Change it to...
percent	$\frac{}{100}$
is	$=$
of, times, product	$\times$
what (or any unknown value)	any variable (x, k, b)
what percent	$\frac{x}{100}$

Here's an example. Try using translation.

▶ What is 30 percent of 200?

First translate, then reduce and solve for the variable, like this.

$$x = 0.30 \times 200$$
$$x = 60$$

That's the answer. 30% of 200 is 60.

Don't be afraid of really big or really small percents, such as $\frac{1}{5}\%$ and 145%. $\frac{1}{5}\%$ just means $\frac{\frac{1}{5}}{100}$, or $\frac{1}{5} \div 100$. That's $\frac{1}{5} \times \frac{1}{100}$, or $\frac{1}{500}$. And 145% just means $\frac{145}{100}$, or 1.45.

Percents and Estimation

Ballparking will come in handy whenever you're asked to estimate percentages. So will that conversion chart we gave you earlier.

Here are some examples:

▶ Estimate 19% of 50.

Use 20% as a friendlier percentage. You can even change that to $\frac{1}{5}$. $\frac{1}{5}$ of 50 = 10.

▶ Estimate 25% of 78.

Use 80 as a friendlier number. $\frac{1}{4}$ of 80 is 20.

▶ Estimate 65% of 34.

Use $\frac{2}{3}$ and 30. $\frac{2}{3}$ of 30 is 20.

▶ Estimate 9.2% of 5.4.

Use 10% and 5. $\frac{1}{10}$ of 5 is $\frac{5}{10}$ or $\frac{1}{2}$ or 0.5.

Percent of Change

To find a percentage increase or decrease, first find the *amount* of increase or decrease, then use the following formula:

$$\text{Amount change} = \frac{x}{100} \times \text{original number (starting point)}$$

Try this example.

▶ Find the percent decrease from 4 to 3.

First figure out what the actual decrease is. The decrease from 4 to 3 is 1. So $1 = \frac{x}{100} \times (4)$, because 4 is the original number. Now solve for x:

$$1 = \frac{4x}{100}$$

$$1 = \frac{x}{25}$$

$$25 = x$$

So the percent decrease from 4 to 3 is 25%.

Some percent questions involve buying and selling items. A **discount** is the amount by which the regular price of an item is reduced.

Here's an example.

▶ Find the sale price of a CD that originally cost $15, but is sold at a 15% discount.

First, find 15% of $15: $(\frac{15}{100})(15)$ = $2.25. That is the discount. Now *subtract* $2.25 from the original price, $15.

$$\begin{array}{r} \$15.00 \\ - \$2.25 \\ \hline \$12.75 \end{array}$$

The sale price of the CD is $12.75.

A **markup** is the difference between the price paid by the merchant and the increased **selling price,** which is the amount the customer pays for the item.

Here's an example.

▶ Find the selling price for a computer for which the store paid $1,300 and the markup is 30%.

First, find 30% of $1,300: $(\frac{30}{100})(\$1,300)$ = $390. That's the markup. Now *add* $390 to the original price of $1,300.

$$\begin{array}{r} \$1,300 \\ + \$390 \\ \hline \$1,690 \end{array}$$

The selling price of the computer is $1,690.

Mile 7: Inequalities, Ratios, and Proportions

INEQUALITIES

Inequalities are sentences that compare quantities. Below are examples of the four inequalities.

6 is greater than 4: $6 > 4$

$2x$ plus 9 is less than 11: $2x + 9 < 11$

x is greater than or equal to 7: $x \geq 7$

11 is less than or equal to $5x$: $11 \leq 5x$

- The sign $<$ means "less than," "fewer than," or "up to," and is represented by an empty circle at the end of a line segment on a number line graph.

- The sign $>$ means "greater than," "more than," "exceeds," or "in excess of," and is represented by an empty circle at the end of a line segment on a number line graph.

- The sign $\leq$ means "less than or equal to," "no more than," or "at most," and is represented by a solid circle at the end of a line segment on a number line graph.

- The sign $\geq$ means "greater than or equal to," "no less than," or "at least," and is represented by a solid circle at the end of a line segment on a number line graph.

For example, the graph of the solution of $5 \leq x < 17$ would look like this.

You can manipulate any inequality in the same way you can an equation, with one important difference. When you multiply or divide both sides of an inequality by a negative number, the direction of the inequality symbol changes. That is, if $x > y$, then $-x < -y$.

Here's an example.

▶ What is x if $12 - 6x > 0$?

You could manipulate this inequality without ever multiplying or dividing by a negative number. Just add $6x$ to both sides. The sign stays the same. Then divide both sides by positive 6. Again, the sign stays the same.

$$12 - 6x > 0$$
$$\underline{+ 6x \quad + 6x}$$
$$12 > 6x$$
$$\frac{12}{6} > \frac{6x}{6}$$
$$2 > x$$

But suppose you subtract 12 from both sides at first.

$$12 - 6x > 0$$
$$\underline{- 12 \qquad - 12}$$
$$-6x > -12$$
$$\frac{-6x}{-6} < \frac{-12}{-6}$$
$$x < 2$$

Notice that the sign flipped that time. You must flip the sign if you divide both sides by a negative number.

RATIOS AND RATES

A **ratio** is a comparison of two numbers by division. It's an abstract relationship in a reduced form. For example, if you were told that there were 14 red marbles and 16 blue marbles in a bowl, the ratio of red to blue marbles in the bowl would be 14:16, or 7:8 (which could also be written as $\frac{7}{8}$). A **rate** is a ratio of two measurements with different units. For example, describing someone's rate of sales, you could say that she sold 24 books in 8 weeks. Books and weeks are different units. Her rate would be 3 books per week.

Say you were told that at a camp for boys and girls, the ratio of girls to boys is 5:3. If the camp's enrollment is 160, how many of the children are boys? A ratio of 5:3 doesn't mean literally 5 girls and 3 boys. It means 8 total "parts" (because 5 + 3 = 8). To find out how many children are in each "part," divide the total enrollment by the number of "parts." Dividing 160 by 8 gives you 20. That means each "part" is 20 children. Three of the "parts" are boys, which means there are 60 boys.

SCALE DRAWINGS AND MODELS

One type of ratio question might involve **scale drawings,** which are used to represent objects that are too large or too small to be drawn or built to actual size. The scale is determined by the ratio of a given length on the drawing or model to its corresponding length in reality. Here's an example.

▶ On a map, the scale is 1 inch = 90 miles. If the distance between Columbus, Ohio, and Chicago, Illinois, on the map is 4 inches, what is the actual distance?

The ratio is $\dfrac{1 \text{ inch}}{90 \text{ miles}}$. Set that equal to $\dfrac{4 \text{ inches}}{x}$ and cross multiply: $1x = (90)(4)$, or 360 miles. So 360 miles is the actual distance.

PROPORTIONS

A **proportion** is an equation that shows that two ratios are equivalent. In a proportion, the **cross products** are equal.

For example, to determine whether $\dfrac{9}{12} = \dfrac{18}{24}$ is a proportion, cross multiply and see if the resulting products are equal. Does $24 \times 9 = 18 \times 12$? Does $216 = 216$? Yes, so $\dfrac{9}{12} = \dfrac{18}{24}$ is a proportion. You also could have reduced both fractions. $\dfrac{9}{12}$ reduces to $\dfrac{3}{4}$, and so does $\dfrac{18}{24}$.
Here's another example.

▶ If 10 baskets contain a total of 50 eggs, how many eggs would 7 baskets contain?

First, set up a proportion, being careful to line up similar units.

$$\frac{\text{Baskets}}{\text{Eggs}} = \frac{\text{Baskets}}{\text{Eggs}}$$

$$\frac{10}{50} = \frac{7}{x}$$

Because you can treat ratios exactly like fractions, you can find x by cross multiplying.

$$10x = (50)(7)$$
$$10x = 350$$
$$x = 35$$

(You could have made the cross multiplication simpler by reducing $\frac{10}{50}$ to $\frac{1}{5}$ before cross multiplying.)

SEQUENCES

A **sequence** is a list of numbers in a certain order. Each number in a sequence is called a **term.** When the difference between any two consecutive terms is the same, the sequence is called an **arithmetic sequence.** The difference between the terms in an arithmetic sequence is called the **common difference.** If the consecutive terms of a sequence are formed by multiplying by a constant factor, it's a **geometric sequence,** and the factor is called a **common ratio.**

For example,

$$2, 4, 6, 8, 10 \ldots$$

is an arithmetic sequence with a common difference of 2.

$$2, 4, 8, 16, 32 \ldots$$

is a geometric sequence with a common ratio of 2.

Here's an example.

▶ Identify the following sequence as either arithmetic, geometric, or neither, and find the next three terms:
1, 2, 5, 10, 17 . . .

First, check if the example is an arithmetic sequence: The difference between 1 and 2 is 1; the difference between 2 and 5 is 3; the difference between 5 and 10 is 5; the difference between 10 and 17 is 7. So this is not an arithmetic sequence. However, it is not a geometric sequence either, because $1 \times 2 = 2$, but $2 \times 2 \neq 5$. (In other words, there's no constant factor.) In this example, each term is 2 greater than the previous term. The next term will be 9 more than 17, or 26. The term after that will be 11 more than 26, or 37. The term after that will be 13 more than 37, or 50.

POWERS AND EXPONENTS

Exponents are a sort of mathematical shorthand. Instead of writing $(2)(2)(2)(2)$ you can write 2^4, which is called a **power.** The little 4 is called an exponent and the big 2 is called a **base.** Here are some rules to remember about exponents.

- Raising a number greater than 1 to a power greater than 1 results in a *bigger* number. For example, $2^2 = 4$.

- Raising a fraction between 0 and 1 to a power greater than 1 results in a *smaller* number. For example, $\left(\frac{1}{2}\right)^2 = \frac{1}{4}$.

- A negative number raised to an even power becomes *positive*. For example, $(-2)^2 = 4$.

- A negative number raised to an odd power remains *negative*. For example, $(-2)^3 = -8$.

- Raising any number to the power of 0 results in 1. For example, $3^0 = 1$.

When you multiply or divide powers, just remember that when you're in doubt, expand it out. In other words, $2^2 \times 2^4$ can be rewritten as $(2 \times 2)(2 \times 2 \times 2 \times 2)$, which is just $2 \times 2 \times 2 \times 2 \times 2 \times 2$, which $= 2^6$. The base is the same, so you can simply add the exponents.

You can use the same method with division: $2^6 \div 2^2 = (2 \times 2 \times 2 \times 2 \times 2 \times 2) \div (2 \times 2) = 2 \times 2 \times 2 \times 2 = 2^4$. The base is kept the same, and you can subtract the exponents.

SCIENTIFIC NOTATION

Large numbers such as 15,000,000 can be rewritten as 1.5×10^7, which is called **scientific notation.** Scientific notation is used to lessen the chance of omitting a zero or misplacing the decimal point. Just move the decimal point over as many spaces as it takes for there to be one unit to the left of it. Then add "$\times$ 10" and raise the 10 to the power that represents the number of times you moved the decimal point. Try one.

▶ Express 595,000,000 in scientific notation.

Move the decimal point over as many spaces as it takes for the number to be 5.95, which is 8. Then add "$\times$ 10" and raise the 10 to the power 8.

The answer is 5.95×10^8. Try another.

▶ Express 2.483×10^5 in standard form.

Now do the opposite: Move the decimal point over the same as the power dictates—
in this case, 5. Add zeros as place holders.
The answer is 248,300.

SQUARE ROOTS

If $x^2 = y$, then x is the square root of y and $\sqrt{y} = x$. A positive square root of a number
is called the **principal square root.** The **radical sign,** $\sqrt{}$, indicates the principal
square root. A negative sign outside the radical sign indicates the negative square
root. For example, $\sqrt{25} = 5$, but $-\sqrt{25} = -5$.

Estimating Square Roots

A **perfect square** is a number with an integer for a square root. For example, 36 is
a perfect square because its square root is 6, but 24 is not a perfect square because
it doesn't have an integer for its square root. This information can help you estimate
square roots.

Say you wanted to estimate the square root of 50. What is the closest perfect square
to 50? 49, with a square root of 7. The next closest perfect square to 50 is 64, with
a square root of 8. So, $\sqrt{50}$ is going to be somewhere between 7 and 8, probably
much closer to 7. Try one.

▶ Estimate $\sqrt{115}$.

Think about perfect squares that are close to 115. $\sqrt{100}$ is 10. $\sqrt{121}$ is 11. Because
115 is between 100 and 121, $\sqrt{115}$ must be between 10 and 11.
The answer is between 10 and 11.

THE REAL NUMBER SYSTEM

An **irrational number** is one that cannot be expressed as $\frac{a}{b}$, where a and b are integers and b doesn't equal 0. The set of rational numbers (numbers that can be expressed as $\frac{a}{b}$, where a and b are integers and b doesn't equal 0) combined with the set of irrational numbers (numbers like $\sqrt{2}$ and $\sqrt{5}$, which don't terminate) make up the set of **real numbers.**

Exercise 3

Try the following problems (answers on page 75):

a. What is 20% of 420?

b. 15 is 30% of what number?

c. If a dress originally priced at $80 is increased to $100, by what percent was the price of the dress increased?

d. If $x + 5 < 2x + 8$, what are the possible values of x?

e. If in a bowl of 72 marbles, the ratio of red to blue marbles is 5:1, how many red marbles are there in the bowl?

f. On a certain map, a distance of 5 miles is represented by 3 inches. If two cities are 90 miles apart, how far apart will they be on the map?

g. How much interest is earned on a $5,000 bank account during a 1-year period, if that account earns 4% simple interest?

h. What is $3^6 \times 3^4$?

i. What is 3.24×10^3?

MILE 8: LINES, ANGLES, AND POLYGONS

Even if you think you know all there is to know about geometry and graphing, read through the rest of the miles anyway. The better you know geometry, the more likely you'll nail the questions on the New York State Grade 8 Math test.

LINES AND ANGLES

Before you get too far on the geometry review, make sure you're familiar with the following geometry terms.

- A **line** (which can be thought of as a perfectly flat angle) is 180 degrees.
- When two lines intersect, four angles are formed, and the sum of the angles is 360 degrees.
- When two lines are **perpendicular** to each other, their intersection forms four 90-degree angles, which are also called **right** angles. (Right angles are identified by little boxes at the intersection of the angle's arms.)
- An **acute** angle is one that measures less than 90 degrees.
- An **obtuse** angle measures more than 90 degrees.
- **Supplementary** angles add up to 180 degrees.
- **Complementary** angles add up to 90 degrees.
- **Congruent** angles have the same measure.
- A **triangle** (three-sided figure) contains 180 degrees.
- Any **quadrilateral** (four-sided figure) contains 360 degrees.
- A **circle** contains 360 degrees.
- A **polygon** is a simple closed figure in a plane formed by three or more line segments.

When lines are **parallel,** they never meet. A **transversal** is a line that intersects two or more other lines.

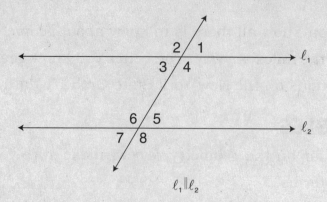

In the figure above, the measures of angles 1, 3, 5, and 7 are equal, and the measures of angles 2, 4, 6, and 8 are equal. Angles 1 and 7, and 2 and 8 are **alternate exterior angles.** Angles 3 and 5, and 4 and 6 are **alternate interior angles.**

Vertical angles are the angles across from each other that are formed by the intersection of lines. Vertical angles are always equal.

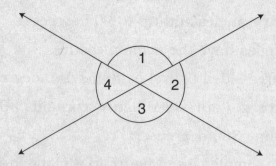

In the figure above, angles 1 and 3 are vertical angles, and are therefore equal. Angles 2 and 4 are vertical angles, and are therefore equal.

CLASSIFYING QUADRILATERALS

A **parallelogram** is a quadrilateral with two pairs of opposite sides that are parallel.

A **rectangle** is a parallelogram with four right angles.

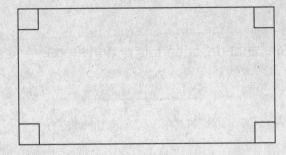

A **rhombus** is a parallelogram with all sides congruent.

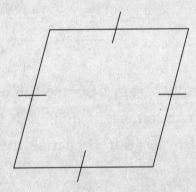

A **square** is a parallelogram with all sides congruent and four right angles.

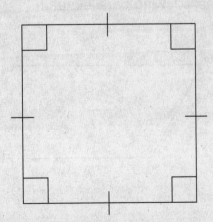

A **trapezoid** is a quadrilateral with exactly one pair of opposite sides that are parallel.

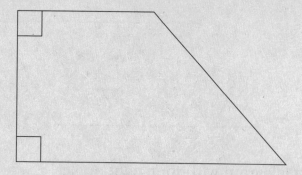

Try this example.

▶ Identify all the names that describe this quadrilateral.

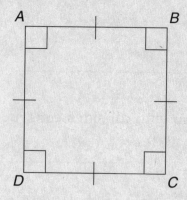

Quadrilateral *ABCD* is a parallelogram because the opposite sides are parallel. It is also a rectangle because it has four right angles. It is also a rhombus because all sides are congruent. It is also a square because the sides are congruent and the angles are all 90 degrees.

Here's another example.

▶ Find the value of *x* in the following parallelogram:

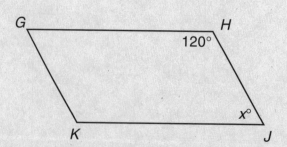

In a parallelogram, the opposite angles are equal. Because the measure of angle H is 120 degrees, the measure of angle K must also be 120 degrees. Because there are 360 degrees in any quadrilateral, the measures of the other two angles, G and J, must add up to 360 – 120 (angle H) – 120 (angle K).

$$360 - 120 - 120 = 120$$

Because the measures of angles G and J must be equal, to find x, which is angle J, you can divide 120 by 2.

$$120 \div 2 = 60$$

So $x = 60$ degrees.

PERIMETER AND AREA

The **perimeter** of a polygon is the sum of the measures of all of its sides. The perimeter of a rectangle is $P = 2l + 2w$.

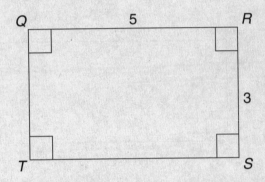

The perimeter of rectangle $QRST$ is 2(5) + 2(3), or 16.

A rectangle with four equal sides is a square, and the perimeter is $P = 4s$.

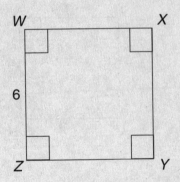

The perimeter of square $WXYZ$ is 4(6), or 24.

In parallelograms, each pair of opposite sides are parallel and have the same length. The formula for the perimeter of a parallelogram is $P = 2a + 2b$, where a and b are adjacent sides.

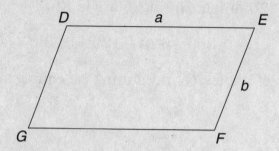

In the figure above, $a = 6$ and $b = 4$. So the perimeter of parallelogram $DEFG$ is $2(6) + 2(4)$, or 20.

Area is the measure of the surface enclosed by the figure. The area of a rectangle is $A = lw$.

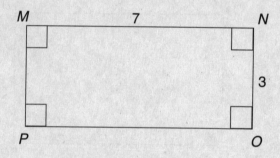

The area of rectangle $MNOP$ is $(7)(3)$, or 21.

The formula for the area of a square is $A = s^2$.

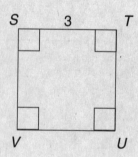

The area of square $STUV$ is 3^3, or 9.

The formula for the area of a parallelogram is $A = bh$, where b is the base and h is the height. The height is really the **altitude,** which is a segment in a polygon that is perpendicular to the base.

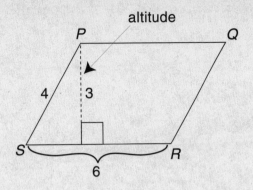

The area of parallelogram *PQRS* is (3)(6), or 18.

The formula for the area of a triangle is $A = \frac{1}{2}bh$.

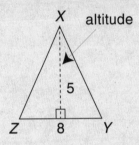

The area of triangle *XYZ* is $\frac{1}{2}(8)(5)$ or 20.

The formula for the area of a trapezoid is $A = \frac{1}{2}h(a + b)$, where a and b are the parallel sides of the trapezoid.

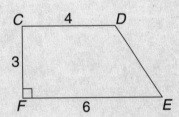

The area of trapezoid *CDEF* is $\frac{1}{2}(3)(4 + 6)$, or $\frac{1}{2}(30)$, or 15.

Tip

Remember that the height of a figure may not be the measure of the tallest line. A figure's height is its altitude, often measured with a dotted line from the base.

CLASSIFYING TRIANGLES

A triangle is a polygon with three sides. It can be classified by its angles or the lengths of its sides.

A **scalene** triangle has no sides of equal length.

An **isosceles** triangle has at least two sides of equal lengths. This means that two of the angles are also equal, and that the third angle does not have to be.

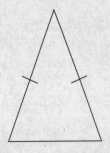

In an **equilateral** triangle, all three sides are of equal length, and all angles equal 60 degrees (because a triangle contains 180 degrees).

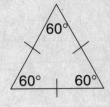

Congruent Triangles

Triangles that have the same shape and size are called **congruent** triangles. The parts of congruent triangles that "match" are called **corresponding parts.** Sometimes you only need to know that three of the six parts are congruent to know that two triangles are congruent.

Two triangles are congruent if the following corresponding parts of the triangles are congruent:

Three sides, or SSS (side-side-side)

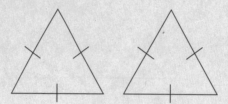

Two angles and the included side, or ASA (angle-side-angle)

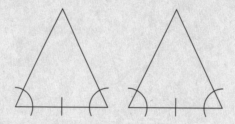

Two sides and the included angle, or SAS (side-angle-side)

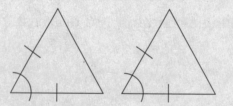

Try this example.

► Determine whether these two triangles are congruent.

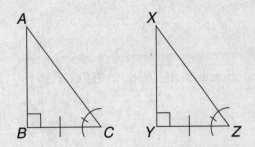

Angles B and Y are both 90-degree angles, BC and YZ are congruent, and angles C and Z are also congruent; therefore triangles ABC and XYZ are congruent by ASA, or angle-side-angle.

Similar Triangles

Triangles that have the same shape, but may differ in size, are called **similar triangles.** The angles are equal, but the sides aren't. The sign "~" indicates similarity.

For example,

► Triangle A has angles measuring 110 degrees, 40 degrees, and a degrees. Triangle B has angles measuring 40 degrees, 30 degrees, and b degrees. Are they similar?

The sum of the measures of angles in a triangle is always 180 degrees. Therefore, the measures of the angles in triangle A are $110 + 40 + a = 180$. That's $150 + a = 180$. So $a = 30$. In triangle B, the measures of the angles $40 + 30 + b = 180$. That's $70 + b = 180$. So $b = 110$.

Now you know that the angles in the triangle A are 110, 40, and 30, and the angles in triangle B are 40, 30, and 110. Because the measures of all three angles in both triangles are equal, triangle A ~ triangle B.

SIMILAR POLYGONS

Two polygons are similar if they are the same shape but different sizes. This means that their corresponding angles are congruent and their corresponding sides are in proportion.

Try this example.

▶ Determine whether rectangles *ABCD* and *EFGH* are similar.

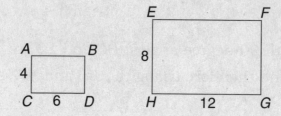

Because both polygons are rectangles, you know that all of the angles are equal. To determine whether the rectangles are similar, you have to see if the corresponding sides are in proportion.

Start with the heights. If *ABCD* and *EFGH* are similar, then $\frac{AC}{EH}$ will equal $\frac{CD}{HG}$. *AC* = 4 and *EH* = 8. So $\frac{AC}{EH}$ is $\frac{4}{8}$, or $\frac{1}{2}$ (simplify whenever possible!). *CD* = 6 and *HG* = 12. So $\frac{CD}{HG}$ is $\frac{6}{12}$, or $\frac{1}{2}$. Because the sides of *ABCD* and *EFGH* are in proportion, you know they are similar.

Exercise 4

Try the following problems (answers on page 76):

a. What are all the missing angles in the diagram below?

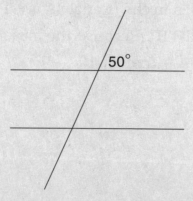

b.   The area of a square with side 6 is how much greater than its perimeter?

c.   What is the area of the square below? What is the area of the triangle?

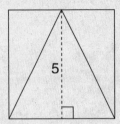

d.   What are the 3 ways to know that two triangles are congruent?

e.   The following two right triangles are congruent. If x = 60 degrees, what is the value of y?

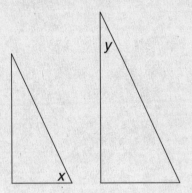

SYMMETRY

A figure has **line symmetry** if it can be folded so that one half of the figure coincides with the other half. A **reflection** is where a figure is flipped over a **line of symmetry** (which divides a figure into two identical halves).

For example, look at the letter **A**. If you split it in half with a vertical line, the two halves would be identical. So the letter **A** has line symmetry. Now look at the letter **Z**. Try splitting it in half with a vertical line, and then a horizontal line. It cannot be split in half to make two identical halves. So the letter **Z** does not have line symmetry.

A **rotation** is when something is turned completely around a central point. A figure has **rotational symmetry** if it looks exactly the same after it has been rotated 360 degrees.

MILE 10: THE PYTHAGOREAN THEOREM, TRIGONOMETRY, CIRCLES, AND SOLIDS

PYTHAGOREAN THEOREM

In a right triangle, the square of the length of the **hypotenuse** (the side opposite the right angle) is equal to the sum of the squares of the lengths of the legs. This is proven by the **Pythagorean theorem,** or $c^2 = a^2 + b^2$, where c is the hypotenuse.

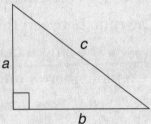

Don't worry, this formula will be on your reference sheet when you take your New York State Grade 8 Math test.

Look at an example.

▶ Determine the length of the hypotenuse in the triangle below.

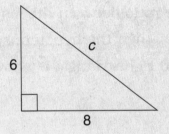

Use the Pythagorean theorem.

$$c^2 = a^2 + b^2$$
$$c^2 = 6^2 + 8^2$$
$$c^2 = 36 + 64$$
$$c^2 = 100$$
$$c = \sqrt{100}$$
$$c = 10$$

The length of the hypotenuse is 10 inches.

So the numbers 6, 8, and 10 work in a right triangle because they satisfy the Pythagorean theorem. Such numbers are called **Pythagorean triples.** The 6-8-10 triple is a multiple of the 3-4-5 triple. Other common Pythagorean triples include the 5-12-13 right triangle and the 8-15-17 right triangle.

Special Right Triangles

There are two special right triangles in which the sides and angles have certain relationships. The first is the 45:45:90, or the **isosceles right triangle.** It's half of a square. In such a triangle, the two legs are equal. If you know one of the sides, you can find the other two sides. The ratio between the length of either of them and that of the hypotenuse is $1:\sqrt{2}$. That is, if the length of each short leg is x, then the length of the hypotenuse is $x\sqrt{2}$. For example, here's an isoceles right triangle.

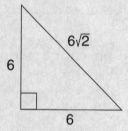

The other special right triangle is the 30:60:90 right triangle. It's half of an equilateral triangle. Again, if you know the length of any of the sides, you can find the lengths of the others. The ratio between the length of the legs is $1:\sqrt{3}:2$. That is, if the shortest side is length x, then the hypotenuse is $2x$ and the remaining side is $x\sqrt{3}$. For example, a 30:60:90 right triangle is shown below.

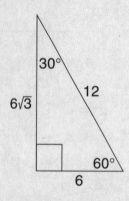

TRIGONOMETRY

Trigonometry is the study of the relationships among the sides of right triangles. There are certain formulas that represent these relationships.

You can remember these trigonometric relationships with the word **SOHCAHTOA**.

$$\text{\textbf{S}ine} = \frac{\textbf{O}pposite}{\textbf{H}ypotenuse}$$

$$\text{\textbf{C}osine} = \frac{\textbf{A}djacent}{\textbf{H}ypotenuse}$$

$$\text{\textbf{T}angent} = \frac{\textbf{O}pposite}{\textbf{A}djacent}$$

Don't worry, these formulas will be on your Reference Sheet when you take your test. For example, here's a right triangle.

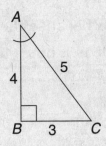

If you were asked to find the sine of the marked angle, angle A, you would put the opposite side, which is 3, over the hypotenuse, which is 5. The sine of angle A is $\frac{3}{5}$.

If you were asked to find the cosine of angle A, you would put the adjacent side, which is 4, over the hypotenuse, which is 5. The cosine of angle A is $\frac{4}{5}$.

If you were asked to find the tangent of angle A, you would put the opposite side, which is 3, over the adjacent side, which is 4. The tangent of angle A is $\frac{3}{4}$.

CIRCLES

When you deal with circles, you have to deal with π, which is the ratio between the circumference of a circle and its diameter. The value of π is approximately 3.14, or $\frac{22}{7}$. You'll be told on your test which version to use.

The circumference of a circle is like the perimeter of a quadrilateral: It's the distance around the outside. The formula for finding the circumference of a circle is π times twice the radius ($C = 2\pi r$), or π times the diameter ($C = \pi d$).

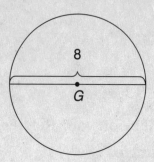

The circumference of circle G is 8π because the diameter is 8.

The formula for the area of a circle is π times the square of the radius ($A = \pi r^2$).

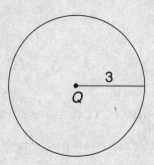

The area of circle Q is 9π because the radius is 3. ($3^2 = 9$.)

THREE-DIMENSIONAL FIGURES

Three-dimensional figures are called **solids. Prisms** have flat surfaces, called **faces.** The faces meet to form **edges.** The edges meet at corners called **vertices. Bases** are used to name the prism. For example, a rectangular prism is a three-dimensional solid with a rectangle for a base. A **pyramid** has a polygon for a base and triangles for sides.

This is a **rectangular prism.**

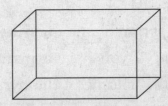

This is a **cylinder.**

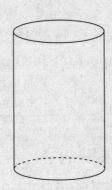

This is a **pyramid.**

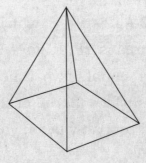

This is a **cone.**

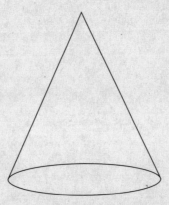

Volume of Solids

Volume is the measure of the space occupied by a solid. The volume of a prism equals the area of base times the height, or $V = Bh$. Measuring the area of the base of a solid depends on what kind of shape it is.

For a rectangular prism, B is equal to lw because the base is a rectangle. So for a rectangular prism, $V = lwh$. For a cylinder, B is equal to πr^2 because the base is a circle. So for a cylinder, $V = \pi r^2 h$.

The relationship between a pyramid and a prism is similar to the relationship between a cylinder and a cone. If you have a pyramid and a prism with the same base and height, the ratio of the volume of the pyramid to the volume of the prism is 1:3. So the volume of a cone $= \frac{1}{3}Bh$ or $\frac{1}{3}\pi r^2 h$, and the volume of a pyramid is $\frac{1}{3}Bh$, or $\frac{1}{3}lwh$.

The formulas for the volumes of some objects will be on your reference sheet when you take your test. Still, many important formulas *won't* be on the reference sheet, so it's helpful to memorize them.

Surface Area of Solids

The **surface area** of a prism is the sum of the area of its faces. If you had a box whose dimensions were 2 by 3 by 4, there would be two sides that are 2 by 3 (area of 6), two sides that are 3 by 4 (area of 12), and two sides that are 2 by 4 (area of 8). So the surface area would be 6 + 6 + 12 + 12 + 8 + 8, which is 52. Therefore, for a rectangular prism, the formula for surface area is $2lh + 2lw + 2wh$.

The surface area of a cylinder equals two times the area of the circular bases $(2\pi r^2)$ + the area of the curved surface $(2\pi rh)$, or $2\pi r^2 + 2\pi rh$.

Don't worry, these formulas will probably be on your Reference Sheet when you take your test.

MILE 11: THE COORDINATE SYSTEM, FUNCTIONS, AND TRANSFORMATIONS

THE COORDINATE SYSTEM

The **coordinate system** is shaped like a grid. The horizontal line is called the *x*-axis; the vertical line is called the *y*-axis. The four areas formed by the intersection of these axes are called **quadrants.** The point where the axes intersect is called the **origin.** This is what it looks like.

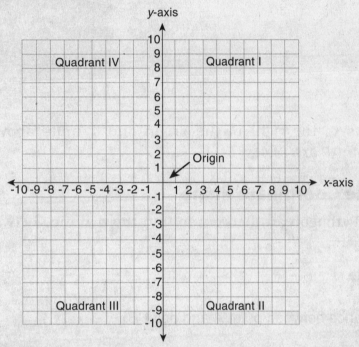

To express any point in the coordinate system, first give the horizontal value, then the vertical value, or (*x,y*). For example, look at the coordinate grid on the next page.

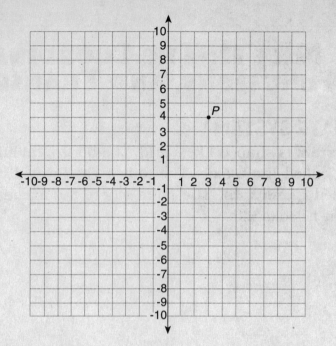

Point P is 3 units to the right of the origin and 4 units above it. Therefore, the coordinates of point P are (3,4).

Distance on the Coordinate Plane

You can use the Pythagorean theorem to find the distance between two points on a grid.

Here's an example.

▶ Find the distance between point A and point B.

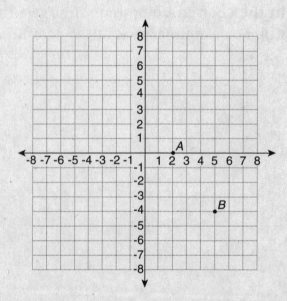

First, draw a line connecting point *A* and point *B*. Then, make it the hypotenuse of a right triangle.

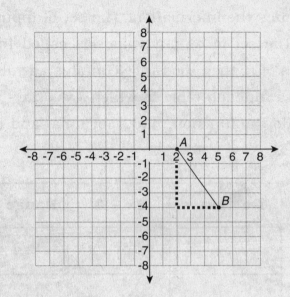

Now, all you have to do is count the boxes to find the length of the two legs of the triangle, and use those numbers in the Pythagorean theorem to solve for the hypotenuse.

$$c^2 = a^2 + b^2$$
$$c^2 = 3^2 + 4^2$$
$$c^2 = 9 + 16$$
$$c^2 = 25$$
$$c = \sqrt{25}$$
$$c = 5$$

The distance from point *A* to point *B* is 5 units. (You might have recognized the 3-4-5 triangle at an earlier stage.)

FUNCTIONS

A relationship in which one thing depends upon another is called a **function;** a **function table** organizes the information. The set of input values in a function is called the **domain.** The set of output values is called the **range.** If the domain contains all the values of *n*, then the range contains all values of *f(n)*.

Here's an example of a function table in which $f(x) = 2x - 4$. Basically, the domain value gets plugged into the function, like a formula, and the answer is the range. Take a look.

Domain *x*	2*x* – 4	Range *f(x)*
–2	2(–2) – 4	–8
–1	2(1) – 4	–6
0	2(0) – 4	–4
1	2(–1) – 4	–2
2	2(2) – 4	0

This equation has an infinite number of solutions; in other words, you could make *x* almost any value, and you'd get another value in the range column. If you were going to graph the information from the function table above, you would graph the points as (domain,range). In other words, (–2,–8), (–1,–6), (0,–4), (1,–2), (2,0). See what it would look like on the next page.

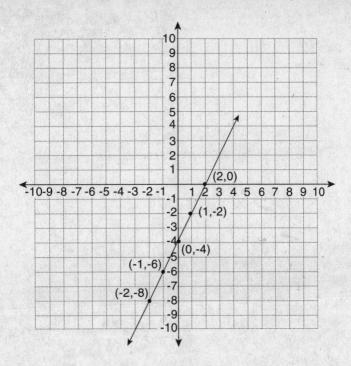

An equation in which the graphs of the solutions form a line is called a **linear function.**

TRANSLATIONS, REFLECTIONS, ROTATIONS

To **translate** a point as described by an ordered pair, add the coordinates of the ordered pair to the coordinates of the point. In other words, (x,y) translated by (a,b) becomes $(x + a, y + b)$. If you had to translate $(3,4)$ by $(2,2)$, the new coordinates would be $(5,6)$.

To **reflect** a point over the x-axis, use the same x-coordinate and multiply the y-coordinate by -1. In other words, (x,y) becomes $(x,-y)$, and $(3,4)$ would become $(3,-4)$. To reflect a point over the y-axis, use the same y-coordinate and multiply the x-coordinate by -1. In other words, (x,y) becomes $(-x,y)$, and $(3,4)$ would become $(-3,4)$.

A **rotation** moves a figure around a central point. To rotate a figure 90 degrees counterclockwise about the origin, switch the coordinates of each point and then multiply the new first coordinate by -1. In other words, (x,y) becomes $(-y,x)$, and $(3,4)$ would become $(-4,3)$. To rotate a figure 180 degrees about the origin, multiply both coordinates of each point by -1. In other words, (x,y) becomes $(-x,-y)$, and $(3,4)$ would become $(-3,-4)$. To rotate a figure 270 degrees counterclockwise (90 degrees clockwise), switch the coordinates of each point and then multiply the new second coordinate by -1. In other words (x,y) becomes $(y,-x)$, and $(3,4)$ would become $(4,-3)$.

GRAPHS THAT DISPLAY DATA

A **bar graph** displays the frequency of data using a series of rectangles (bars). The information on a bar graph can be obtained from a **frequency table,** which displays both axes of the bar graph.

This is what a bar graph looks like.

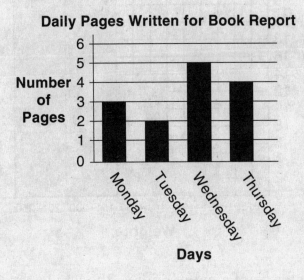

This is what a frequency table can look like.

Interval	Frequency	Cumulative Frequency
51–100	7	7
101–150	7	14
151–200	10	24
201–250	6	30

A **histogram** is a special type of bar graph that displays data in intervals of equal size. The intervals cover all possible values of data; therefore, there are never any spaces between the bars of a histogram. Below is an example of a histogram based on the data in the frequency table on the previous page.

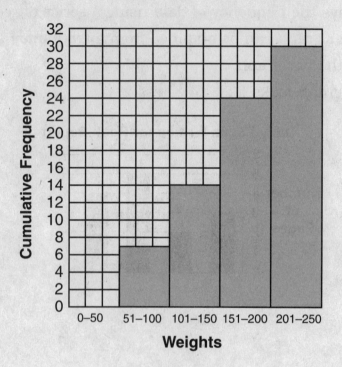

A **circle graph** compares parts of a set of data to the whole set. This is what it looks like.

Studying and making observations about data is called **data analysis.**

MEASURES OF CENTRAL TENDENCY

Numbers or pieces of data that can represent the whole set of data are called **measures of central tendency.** They are the **mean,** the **mode,** and the **median.** The mean is the sum of the numbers in the set of data divided by the number of pieces of data (this is sometimes called the **average** or **arithmetic mean**); think of it as $A = \dfrac{T}{N}$, where A = average, T = total, and N = the number of things. The mode is the number or piece of data that appears the most often (remember *mode = most*). The median is the number in the middle of a set of data when the data are arranged in order (remember *median = middle*). When you have an even amount of numbers, there will be two middle numbers, and the median will be the mean of those two numbers. All you have to do is add the two middle numbers and divide by 2 to get the median.

PROBABILITY OF SIMPLE EVENTS

An **event** is a specific outcome or type of outcome. **Probability** is the chance that an event will happen. It is the number of ways that an event can occur divided by the number of possible outcomes.

$$P = \frac{\text{number of ways event can occur}}{\text{number of possible outcomes}}$$

If you flip a fair coin, what's the probability that it will land on "heads"? One out of two, or $\dfrac{1}{2}$, because the number of possible outcomes is two, heads and tails. Here are some rules of probability.

- If it is impossible for something to happen, the probability of it happening is equal to 0.

- If something is certain to happen, the probability is equal to 1.

- If it is possible for something to happen, but not necessary, the probability is between 0 and 1, otherwise known as a fraction.

Here's an example.

▶ A bowl contains 5 yellow marbles and 4 green marbles. If 1 marble is picked from the bowl at random, what is the probability that it will be green?

Remember the formula: $P = \dfrac{\text{number of ways event can occur}}{\text{number of possible outcomes}}$

In this case, $\dfrac{\text{number of green marbles}}{\text{total number of marbles}}$, or $\dfrac{4}{9}$.

The answer is $\dfrac{4}{9}$.

Counting Outcomes

A **tree diagram** can be used to find the number of possible outcomes.

Here's an example.

▶ Bobby is choosing a breakfast of eggs, pancakes, and juice from a menu. The eggs are either fried or scrambled. Pancakes can be plain, blueberry, or chocolate chip. Juice is a choice of orange juice or grapefruit juice. How many different breakfasts can be made using one type of egg dish, one type of pancake, and one type of juice?

See what a tree diagram for this question would look like on the next page.

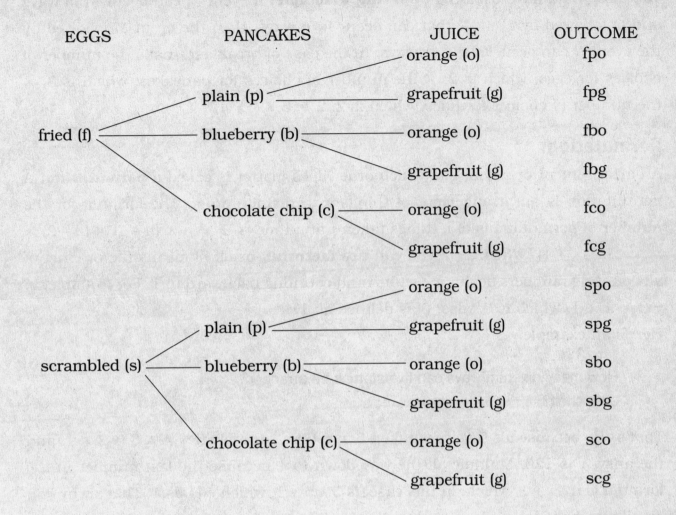

EGGS	PANCAKES	JUICE	OUTCOME

There are twelve different breakfasts, or outcomes. The list of all possible outcomes is called the **sample space.** The sample space for the above breakfast problem is

fpo	spo
fpg	spg
fbo	sbo
fbg	sbg
fco	sco
fcg	scg

You also could have used the **Counting Principle:** If event M can occur in m ways and is followed by event N that can occur in n ways, then the event M followed by the event N can occur in $m \times n$ ways. In the case of breakfast, that's the number of choices for eggs, which is 2, $\times$ the number of choices for pancakes, which is 3, $\times$ the number of choices for juice, which is 2. $2 \times 3 \times 2 = 12$.

Permutations

An arrangement or a listing in which order does matter is called a **permutation.** A permutation is an arrangement of things in a definite order. The formula for the number of permutations of n things taken r at a time is $_nP_r = n \times (n - 1) \times (n - 2) \times \ldots \times (n - r + 1)$. When $n = r$, you can use **factorial,** or $n!$. $n!$ means the product of all counting numbers beginning with n and counting backward to 1. For example, $3! = 3 \times 2 \times 1$, which is 6. Also, $0!$ is defined as 1.

Here's an example.

▶ How many different ways can five statues be arranged on a shelf?

That's $_5P_5$, because it's 5 statues taken 5 at a time. Multiply $5 \times 4 \times 3 \times 2 \times 1$, and the answer is 120. Multiply all the way down to 1 because the last number in the formula is $n - r + 1$, which, in this case, is $5 - 5 + 1$, which equals 1. That's why you could also have called this $5!$, which is also $5 \times 4 \times 3 \times 2 \times 1$, or 120.

Here's another one.

▶ How many ways can you display seven books in groups of three?

In this case, it's $_7P_3$, because it's 7 books taken 3 at a time. Using $n \times (n - 1) \times (n - 2) \times \ldots \times (n - r + 1)$, that's $7 \times 6 \times 5$, which is 210. You must stop at 5 because the last number in the formula is $n - r + 1$, which, in this case, is $7 - 3 + 1$, or 5.

Combinations

An arrangement or a listing in which order does *not* matter is called a **combination.** The formula $_nC_r = \dfrac{n!}{(n - r)!\,(r!)}$ means the number of combinations of n things taken r at a time. Look at the example on the next page.

► How many different 3-flavor combinations of ice-cream flavors could Rachel make out of the following five flavors: vanilla, chocolate, strawberry, butter pecan, and mocha?

In this case, order doesn't matter, because vanilla-chocolate-strawberry is the same as chocolate-strawberry-vanilla. You could use the counting principle, making sure not to repeat orders.

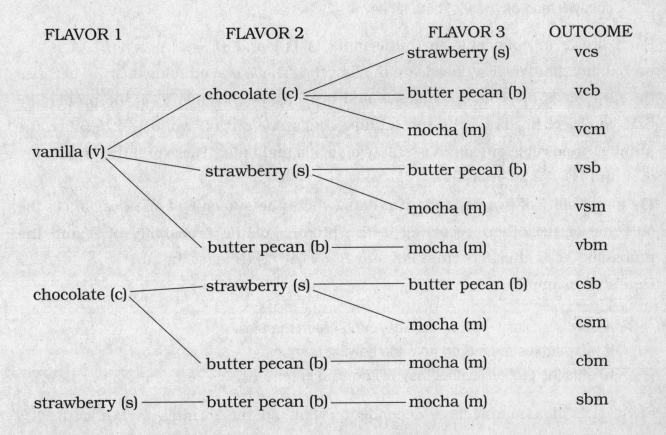

FLAVOR 1	FLAVOR 2	FLAVOR 3	OUTCOME
		strawberry (s)	vcs
	chocolate (c)	butter pecan (b)	vcb
		mocha (m)	vcm
vanilla (v)	strawberry (s)	butter pecan (b)	vsb
		mocha (m)	vsm
	butter pecan (b)	mocha (m)	vbm
chocolate (c)	strawberry (s)	butter pecan (b)	csb
		mocha (m)	csm
	butter pecan (b)	mocha (m)	cbm
strawberry (s)	butter pecan (b)	mocha (m)	sbm

That's 10 combinations. Or you could use the formula $_nC_r = \dfrac{n!}{(n-r)!\,(r!)}$. This problem gives you 5 flavors being used 3 at a time. Plug that into the formula.

$$\frac{5!}{(5-3)!\,(3!)}$$

$$\frac{5 \times 4 \times 3 \times 2 \times 1}{(2 \times 1)(3 \times 2 \times 1)}$$

Don't forget to reduce.

$$\frac{5 \times 4}{2 \times 1} = \frac{20}{2} = 10.$$

The answer is 10 combinations.

PROBABILITY OF COMPOUND EVENTS

Compound events consist of two or more events. The probability of two **independent** events (where the outcome of one doesn't affect the outcome of the other) can be found by multiplying the probability of the first event by the probability of the second event. The formula is P(A and B) = P(A) $\times$ P(B).

▶ If you flip a fair coin twice, what's the probability that the coin will land on "heads" both times?

Each flip is independent, so the formula is P(A and B) = P(A) $\times$ P(B). A is the probability that you'll get heads on the first flip. That's one out of two, or $\frac{1}{2}$, because the number of possible outcomes is still two, heads and tails. B is the probability that you'll get heads on the second flip. That's one out of two, or $\frac{1}{2}$, because the number of possible outcomes is still two, heads and tails. Therefore, the probability of A and B = $\frac{1}{2} \times \frac{1}{2}$, or $\frac{1}{4}$.

The probability of two **dependent** events (where the outcome of one *does* affect the outcome of the other) occurring is the product of the probability of A and the probability of B after A occurs. P(A and B) = P(A) $\times$ P(B following A).

Here's an example.

▶ A bowl contains 5 yellow marbles and 5 green marbles. If two marbles are picked from the bowl at random, what is the probability that they will *both* be green?

Each pick of a marble is a dependent event, so the formula is P(A and B) = P(A) $\times$ P(B following A). Figure out the probability one draw at a time. On the first draw, the probability of drawing a green marble is 5 out of 10, or $\frac{1}{2}$. But now that marble is no longer in the bowl. So on the second draw, the probability of drawing a green marble is 4 out of 9. Therefore, the probability of both marbles being green is equal to $\frac{1}{2} \times \frac{4}{9}$, which is $\frac{4}{18}$, or $\frac{2}{9}$. The answer is $\frac{2}{9}$.

Exercise 5

Try the following problems (answers on page 76):

a. What is the hypotenuse of each of the following triangles?

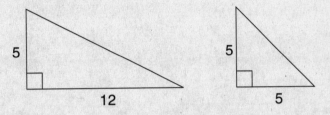

b. What is the tangent of angle A in the triangle below?

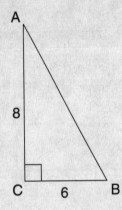

c. What is the circumference and area of a triangle with radius 4?

d. What is the volume of a cube with side 3?

e. If a function f(x) = 3x – 5, what is the value of the function when x is 0? When x is 5?

f. If Jason scores 80, 88, and 90 on his history tests, what is his average on these tests?

g. If Annie has a bowl of 8 red marbles and 10 blue marbles, and she draws one marble at random, what is the chance that she will draw a red marble?

ANSWERS TO REVIEW EXERCISES

Answers to Exercise 1 (from page 16)

a. 11   b. 16
c. 96   d. 30
e. 624   f. 1, 64, 2, 32, 4, 16, 8, 8

g. 2, 3 (all of the factors are 1, 48, 224, 316, 412, and 68, but only 2 and 3 are prime)

Answers to Exercise 2 (from page 23)

a. 8   b. 12   c. $\frac{16}{16}$, which equals 1.

d. $\frac{5}{12}$ e. $\frac{6}{24}$, which equals $\frac{1}{4}$.

f. 2   g. $x = 3$ h. $x = -1$

Answers to Exercise 3 (from page 37)

a. Translate 20% of 420 as $\frac{20}{100} \times 420 = 84$

b. Translate 15 equals 30% of what number as: $15 = \frac{30}{100}x$, so $x = 50$.

c. The increase from 80 to 100 is an increase of 20. This over the original price gives us $\frac{20}{80}$ which is the same as $\frac{1}{4}$, or 25%.

d. $x > -3$

e. In the ratio of 5:1, the total number of parts is 6. Because the actual total number of marbles is 72, and $\frac{72}{6} = 12$, you will need to multiply everything by a factor of 12. Therefore, the actual number of red marbles is 5×12, or 60, and the number of blue marbles is 1×12, or 12. This makes a total of 72 marbles.

f. Set up the proportion $\frac{5 \text{ miles}}{3 \text{ inches}} = \frac{90 \text{ miles}}{x}$. Solve for x, and you will get $x = 54$.

g. $200

h. 3^{10}

i. 3,240

Answers to Exercise 4 (from page 50)

a.

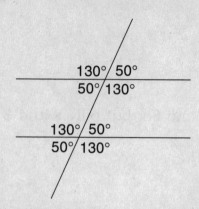

b. The area of a square with side 6 is 6(6), or 36. Its perimeter is (4)(6), or 24. So its area is 12 more than its perimeter.

c. Because the square has side 5, its area is 25. The triangle has height 5 and base 5, so its area is $\frac{1}{2}$(5)(5), or 12.5.

d. Angle-Side-Angle (ASA), Side-Angle-Side (SAS), or Side-Side-Side (SSS).

e. Because both triangles are congruent, both have a right angle (which is 90 degrees), and both have an angle which is 60 degrees, they must have a third angle which measures 30 degrees. So $y = 30$ degrees.

Answers to Exercise 5 (from page 73)

a. 13 and $5\sqrt{2}$

b. $\frac{6}{8}$ or $\frac{3}{4}$

c. 8π and 16π

d. 27

e. –5 and 10.

f. 86

g. $\frac{8}{18}$, or $\frac{4}{9}$

PART III:
THE PRACTICE
TESTS

HOW TO TAKE THE PRACTICE TESTS

Congratulations! You made it through the eighth-grade math review. We hope you used the review in conjunction with your textbook, and if you found something that you weren't clear on, you looked it up in your textbook and reviewed it.

Now that you've reviewed the material, take another look at Miles 1 and 2 to review the test-taking techniques presented there.

Once you've done that, you can move on to the Practice Tests. These Practice Tests contain questions that are meant to approximate the kinds of questions you'll probably see on the Grade 8 Math test.

Detach each bubble sheet from the book before you take each test. Mark your answer to the questions in part 1 on your bubble sheet. The bubble sheet for Practice Test One is on page 83; the bubble sheet for Practice Test Two is on page 85.

Take Practice Test One and see how you do. Time yourself, or have someone time you, as if it were the real test. Give yourself about 35 minutes for Session 1, Part 1, about 35 minutes for Session 1, Part 2, and about 70 minutes for Session 2. Don't forget to detach your Tools, which can be found on the following page, and Reference Sheet, which can be found on page 81.

The answers and explanations follow each test, with the questions reprinted in case you want to make notes. The multiple-choice questions are easily checked, because they only have one right answer. But because the extended-response questions are graded with varying numbers of points by teachers who are determining partial credit, you can't really score your practice tests (unless you have a bunch of eighth-grade math teachers living in your basement!). So use the guidelines in the explanations of those questions to determine how close you got to what would be considered a complete and correct answer.

After you determine whatever questions you got wrong on Practice Test One, go back to the math review and study those topics again, or if necessary, go back to your textbook and review the topics there.

When you are ready, take Practice Test Two and see how you do. After you determine whatever questions you got wrong, go back and study those topics again.

Good luck!

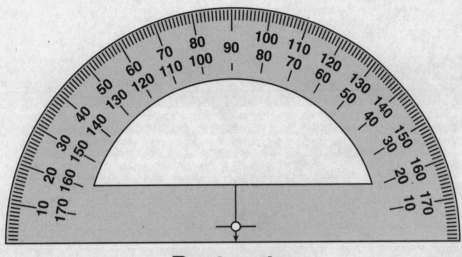

Protractor

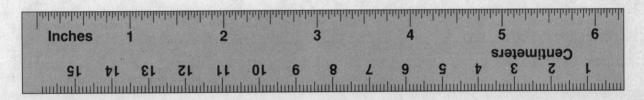

Ruler

MATHEMATICS REFERENCE SHEET

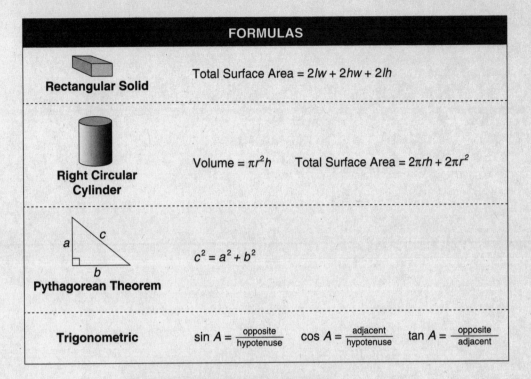

FORMULAS

Rectangular Solid
Total Surface Area = $2lw + 2hw + 2lh$

Right Circular Cylinder
Volume = $\pi r^2 h$ Total Surface Area = $2\pi rh + 2\pi r^2$

Pythagorean Theorem
$c^2 = a^2 + b^2$

Trigonometric
$\sin A = \dfrac{\text{opposite}}{\text{hypotenuse}}$ $\cos A = \dfrac{\text{adjacent}}{\text{hypotenuse}}$ $\tan A = \dfrac{\text{opposite}}{\text{adjacent}}$

TRIGONOMETRIC TABLE

Degrees	Sine	Cosine	Tangent
0	.0000	1.0000	.0000
5	.0872	.9962	.0875
10	.1736	.9848	.1763
15	.2588	.9659	.2679
20	.3420	.9397	.3640
25	.4226	.9063	.4663
30	.5000	.8660	.5774
35	.5736	.8192	.7002
40	.6428	.7660	.8391
45	.7071	.7071	1.0000
50	.7660	.6428	1.1918
55	.8192	.5736	1.4281
60	.8660	.5000	1.7321
65	.9063	.4226	2.1445
70	.9397	.3420	2.7475
75	.9659	.2588	3.7321
80	.9848	.1736	5.6713
85	.9948	.0872	11.4301
90	1.0000	.0000	———

Completely darken bubbles with a No. 2 pencil. If you make a mistake, be sure to erase mark completely. Erase all stray marks.

1. YOUR NAME: _____
   (Print)   Last   First   M.I.

SIGNATURE: _____ **DATE:** _____ / _____ / _____

HOME ADDRESS: _____
(Print)   Number

_____
           City   State   Zip Code

PHONE NO.: _____
(Print)

The **Princeton Review**

Practice Test (1)

Session 1, Part 1	Session 1, Part 1	Session 1, Part 2	Session 2
1. (A) (B) (C) (D)	19. (A) (B) (C) (D)	**Use space provided in your test.**	**Use space provided in your test.**
2. (F) (G) (H) (J)	20. (F) (G) (H) (J)		
3. (A) (B) (C) (D)	21. (A) (B) (C) (D)		
4. (F) (G) (H) (J)	22. (F) (G) (H) (J)		
5. (A) (B) (C) (D)	23. (A) (B) (C) (D)		
6. (F) (G) (H) (J)	24. (F) (G) (H) (J)		
7. (A) (B) (C) (D)	25. (A) (B) (C) (D)		
8. (F) (G) (H) (J)	26. (F) (G) (H) (J)		
9. (A) (B) (C) (D)	27. (A) (B) (C) (D)		
10. (F) (G) (H) (J)			
11. (A) (B) (C) (D)			
12. (F) (G) (H) (J)			
13. (A) (B) (C) (D)			
14. (F) (G) (H) (J)			
15. (A) (B) (C) (D)			
16. (F) (G) (H) (J)			
17. (A) (B) (C) (D)			
18. (F) (G) (H) (J)			

Completely darken bubbles with a No. 2 pencil. If you make a mistake, be sure to erase mark completely. Erase all stray marks.

1. YOUR NAME: _____
   (Print)   Last   First   M.I.

SIGNATURE: _____   DATE: _____ / ____ / ____

HOME ADDRESS: _____
(Print)   Number

_____
                      City   State   Zip Code

PHONE NO.: _____
(Print)

The Princeton Review

Practice Test ②

Session 1, Part 1

1. Ⓐ Ⓑ Ⓒ Ⓓ
2. Ⓕ Ⓖ Ⓗ Ⓙ
3. Ⓐ Ⓑ Ⓒ Ⓓ
4. Ⓕ Ⓖ Ⓗ Ⓙ
5. Ⓐ Ⓑ Ⓒ Ⓓ
6. Ⓕ Ⓖ Ⓗ Ⓙ
7. Ⓐ Ⓑ Ⓒ Ⓓ
8. Ⓕ Ⓖ Ⓗ Ⓙ
9. Ⓐ Ⓑ Ⓒ Ⓓ
10. Ⓕ Ⓖ Ⓗ Ⓙ
11. Ⓐ Ⓑ Ⓒ Ⓓ
12. Ⓕ Ⓖ Ⓗ Ⓙ
13. Ⓐ Ⓑ Ⓒ Ⓓ
14. Ⓕ Ⓖ Ⓗ Ⓙ
15. Ⓐ Ⓑ Ⓒ Ⓓ
16. Ⓕ Ⓖ Ⓗ Ⓙ
17. Ⓐ Ⓑ Ⓒ Ⓓ
18. Ⓕ Ⓖ Ⓗ Ⓙ

Session 1, Part 1

19. Ⓐ Ⓑ Ⓒ Ⓓ
20. Ⓕ Ⓖ Ⓗ Ⓙ
21. Ⓐ Ⓑ Ⓒ Ⓓ
22. Ⓕ Ⓖ Ⓗ Ⓙ
23. Ⓐ Ⓑ Ⓒ Ⓓ
24. Ⓕ Ⓖ Ⓗ Ⓙ
25. Ⓐ Ⓑ Ⓒ Ⓓ
26. Ⓕ Ⓖ Ⓗ Ⓙ
27. Ⓐ Ⓑ Ⓒ Ⓓ

Session 1, Part 2

Use space provided
in your test.

Session 2

Use space provided
in your test.

PRACTICE TEST ONE

SESSION I, PART I

1 Which operation should be performed first in order to solve the following equation?

$$5(10 + 2) - 7 \div 3 =$$

A addition

B subtraction

C multiplication

D division

2 Which one of these represents the prime factorization of 36?

F 4×3^2

G $2^2 \times 3^2$

H $2^3 \times 3^2$

J 6^2

3 A wooden ruler is marked to indicate the lengths $\frac{1}{2}$, $\frac{2}{3}$, $\frac{1}{3}$, and $\frac{1}{5}$ of an inch. Which of these shows the order of these marks, from the largest length to the smallest?

A $\frac{1}{2}, \frac{1}{3}, \frac{1}{5}, \frac{2}{3}$

B $\frac{1}{2}, \frac{1}{3}, \frac{2}{3}, \frac{1}{5}$

C $\frac{1}{5}, \frac{2}{3}, \frac{1}{3}, \frac{1}{2}$

D $\frac{2}{3}, \frac{1}{2}, \frac{1}{3}, \frac{1}{5}$

4 To make the following equation true, which number goes in the box?

$$5^{\square} = 625$$

F 2

G 3

H 4

J 5

5 Kathy and Janine sold some brownies on Saturday. Kathy sold $\frac{1}{4}$ of the brownies, and Janine sold 40% of the brownies. What percentage of the brownies did Kathy and Janine sell altogether?

A 44%

B 50%

C 65%

D 75%

6   The table at a board meeting is shaped like this.

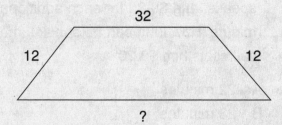

The table is known to have a perimeter of 95 inches. How long is the longest side of the table if the side opposite it is 32 inches long, and the other sides are each 12 inches?

F   12
G   24
H   32
J   39

7   Nikolas ate $\frac{4}{5}$ of his sandwich. What percentage of his sandwich did Nikolas eat?

A   4.5%
B   20%
C   45%
D   80%

8   Use your ruler to solve this problem.

A rectangular pool is shown below.

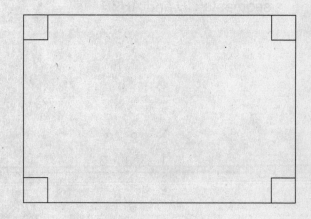

scale: 1 inch = 1 yard

Based on the scale, what is the perimeter of the pool?

F   9 yards
G   10 yards
H   10.5 yards
J   11 yards

Session 1, Part 1

Go to next page

9 In the diagram below, trace over the line segments *AB* and *CD*. Which angle is formed by the intersection of line segments *CD* and *AB*?

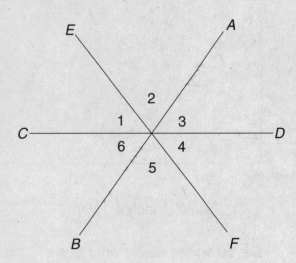

A 1
B 2
C 3
D 4

10 Juan and Simone are buying fruit, and each is looking at an apple. Juan says, "I'll have 5 pieces of fruit if I buy this apple and two more." Simone says, "And I'll have 4 pieces of fruit if I buy this apple and one more." Which of these is correct?

F Juan has 2 pieces of fruit.
G Simone has 3 pieces of fruit.
H Juan has 3 pieces of fruit.
J Simone has 4 pieces of fruit.

11 Leon is figuring out how much a long-distance call will cost him. He gets charged $0.50 for the first minute he speaks, and $0.35 for each additional minute. How long can he speak if the call costs him $8.20?

A 22 minutes
B 23 minutes
C 25 minutes
D 26 minutes

12 Everett is building a fence around his yard that will look like the figure below.

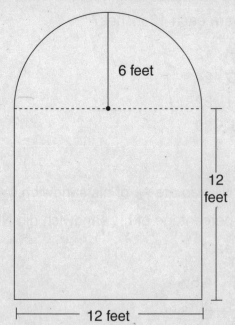

How much material does he need for the fence?

(use π = 3.14)

F 50.68 feet
G 54.84 feet
H 66.84 feet
J 73.68 feet

Session 1, Part 1

Go to next page

The Ice Cream Parlor

Directions: Numbers 13 through 16 are about Becca and her job at the ice cream parlor.

13 On Wednesdays at the ice cream parlor where Becca works, there's a special deal on free toppings, as shown in the table below.

Free Topping Wednesday	
Number of Scoops	Number of Free Toppings
1	2
2	5
3	7
4	10
5	12

If Ms. Meyers receives 17 free toppings with the ice cream she buys on Wednesday, and the pattern continues, how many scoops of ice cream did she buy?

A 6
B 7
C 8
D 9

14 Becca works at the ice cream parlor Monday through Friday, four hours each day, and gets paid $6.50 an hour. How much money has she made after three weeks?

F $130
G $162.50
H $390
J $546

15 Becca is keeping track of the number of ice-cream cones of each flavor she sells this week in the following table:

Weekly Cone Sales	
Flavor	Number of cones sold
chocolate	10
vanilla	12
strawberry	4
coffee	5
pistachio	0
rocky road	6
fudge almond	8
pralines and cream	5

What is the *mean* number of cones of each flavor Becca sold this week?

A 6.25
B 6.5
C 7
D 7.5

Session 1, Part 1

Go to next page

16 Becca's friend Joey comes into the ice cream parlor every Friday to buy a pint of ice cream. This Friday, there are 14 pints of rocky road and 12 pints of fudge almond in the freezer, and no other flavors. If he chooses a pint at random, what is the probability that Joey will buy a pint of rocky road?

F $\frac{7}{13}$

G $\frac{6}{13}$

H $\frac{6}{7}$

J $\frac{7}{6}$

17 In right triangle *EFG* below, which of the following is a true statement?

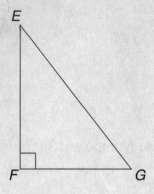

A The sum of angles *E* and *F* is 90 degrees.

B The sum of angles *F* and *G* is 180 degrees.

C The sum of angles *E* and *G* is 90 degrees.

D The sum of angles *E* and *F* is 180 degrees.

18 Which value for *x* will make the statement below true?

$$2 + 4(x + 2) = 14$$

F −1

G 0

H 1

J 2

19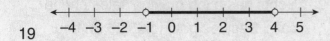

Which of these inequalities is represented on the number line above?

A   −1 < x < 4

B   −1 ≤ x ≤ 4

C   −1 ≥ x > 4

D   −1 < x ≤ 4

20   During her seven-hour shift at the video store, Keesha has to rewind videotapes. She must keep track of the number of videotapes she rewinds each hour.

Videotape Rewinding	
Hour	Numbers of Tapes
1	42
2	35
3	48
4	32
5	29
6	51
7	41

What is the *median* number of tapes Keesha rewound during her seven-hour shift?

F   32

G   39

H   41

J   48

21   Ling has a small six-sided block. Two of the faces are painted red, one is painted blue, one is painted yellow, one is painted green, and one is painted orange. If Ling throws the block up in the air, what is the probability that it will land with the green side facing up?

A $\frac{1}{25}$

B $\frac{1}{6}$

C $\frac{1}{5}$

D $\frac{1}{4}$

22   A recipe calls for 2 eggs for every 3 cups of flour. Which of these proportions shows the way to find how many eggs, *e*, are needed for a recipe calling for 27 cups of flour?

F $\frac{e}{27} = \frac{3}{2}$

G $\frac{27}{e} = \frac{2}{3}$

H $\frac{e}{27} = \frac{2}{3}$

J $\frac{27}{3} = \frac{2}{e}$

23 Which of the following shows a pair of figures that are **similar** but not **congruent?**

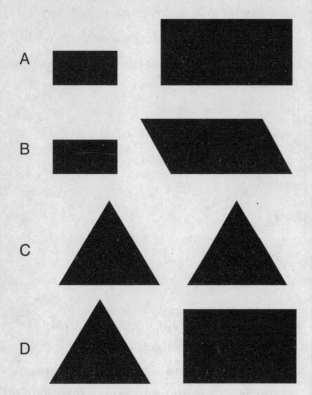

A

B

C

D

24 Maria buys a bag of jelly beans to share with her friend Pierre. Each bag contains two black jelly beans, which Maria always gives to Pierre. If *m* equals the number of jelly beans in a bag, and *p* equals the number after Maria gives Pierre the two black jelly beans, which of the equations below would you use to figure out how many jelly beans are left in the bag after Maria gives Pierre the black ones?

F $p = 2 + m$
G $p - m = 2$
H $p = m - 2$
J $m = p - 2$

25 Bobbi has a job making sales phone calls. Besides getting $5.25 an hour, she also receives an additional $0.30 for each phone call she makes. On Tuesday, Bobbi earned $18.45 and made an average of three phone calls per hour. How many hours did she work on Tuesday?

A 2 hours
B 3 hours
C 4 hours
D 4.5 hours

26 If Kevin can guess how many beans are in a jar, he'll win a prize. He knows the following: There are only red, orange, yellow, and green beans in the jar. The jar contains twice as many orange beans as red beans, three times as many green beans as yellow beans, and half as many yellow beans as red beans. If there are 6 red beans, how many beans are in the jar?

F 20
G 30
H 35
J 40

Session 1, Part 1

Go to next page

27 On the grid below, one triangle is shown. Two coordinates for another triangle are also shown.

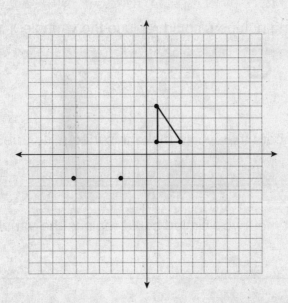

Which of these sets of coordinates will complete a second triangle that is similar to the triangle shown?

A   (−2,−6)

B   (−2,−8)

C   (−2,−9)

D   (−2,−7)

SESSION I, PART 2

28 Which of the following numbers could replace the variable x in the inequality below?

$$0.67 > x > \frac{2}{5}$$

Circle all of the numbers that would make this inequality true.

$\frac{1}{3}$ 0.42 $\frac{6}{8}$ $\frac{1}{2}$ $\frac{3}{2}$

Explain why each number you circled could replace the variable x.

Session 1, Part 2

Go to next page

$$m \div k = k$$

$$n + m = j$$

$$j \times n = j$$

$$n + n = k$$

In the three equations above, variables *j, k, m,* and *n* each represent a different whole number. If *k* = 2, find the value for each of the remaining variables. Show all of your work.

Show your work.

Answers

j = _____

m = _____

n = _____

Session 1, Part 2

Go to next page

30 The members of the marching band are selling T-shirts and sweatshirts to raise money for new uniforms. The following table shows the number of T-shirts and sweatshirts the band sells over a four-day period:

T-Shirt and Sweatshirt Sales		
Day	Number of T-Shirts	Number of Sweatshirts
Monday	42	21
Tuesday	24	28
Wednesday	30	32
Thursday	38	25

Part A

Construct a double line graph of the information from the table on the grid below, being sure to

- title the graph and label the axes
- be accurate
- make a key
- be consistent with the scale

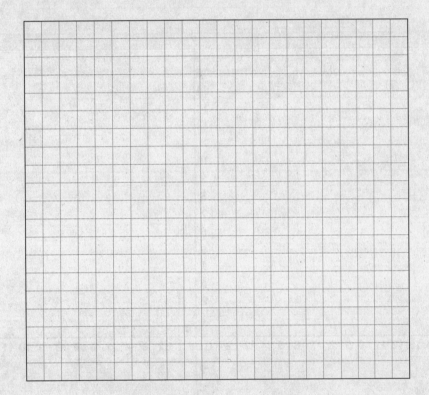

Part B

Using your double line graph, determine on which day the number of T-shirts sold and the number of sweatshirts sold was the **closest.**

Day _____

31 These are the coordinates for quadrilaterals A, B, and C.

Quadrilateral A: (5,4), (8,4), (5,9), (8,11)

Quadrilateral B: (−4,−1), (−4,2), (−9,−1), (−9,2)

Quadrilateral C: (3,−2), (5,−5), (10,−2), (12,−5)

Part A

On the following grid, plot, connect, and label the coordinates of quadrilaterals A, B, and C.

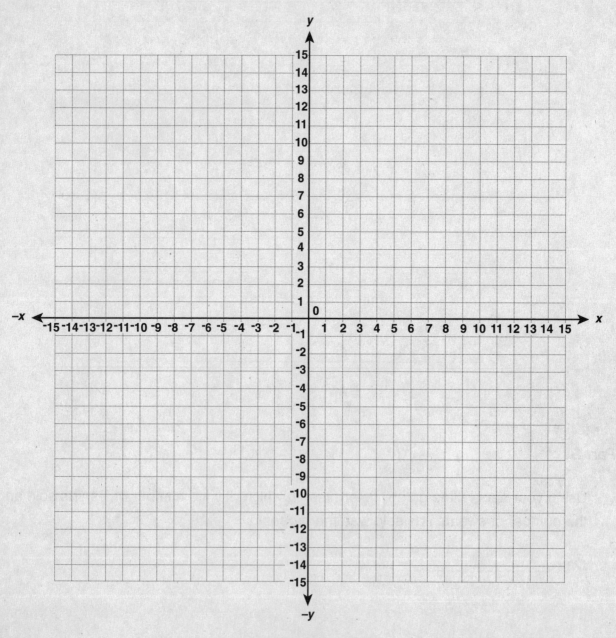

Name and describe what type(s) of quadrilaterals each one is, and explain in words those qualities that helped you identify them.

Quadrilateral A is a _____

_____

_____

_____

Quadrilateral B is a _____

_____

_____

_____

Quadrilateral C is a _____

_____

_____

_____

Session 1, Part 2

Go to next page

32 Vanessa is deciding which outfit to wear to school today. Here is what she has to choose from.

Today's Possible Outfits		
Top	**Bottom**	**Shoes**
Green	Jeans	Sneakers
Red	Skirt	Boots
White		

How many different outfits, each consisting of one top, one bottom, and one pair of shoes, can Vanessa choose?

Show your work.

Answer _____

If Vanessa's friend Jamie randomly selects Vanessa's outfit for her, what is the probability that the jeans and the sneakers will be selected?

Probability _____

Session 1, Part 2

Go to next page

33 The pictured ladder is 10 feet long. The top of the ladder hits the wall 6 feet off the floor. How many feet is the bottom of the ladder from the wall?

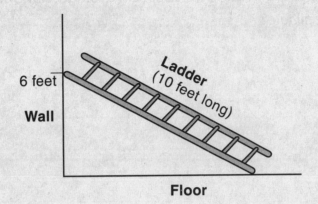

6 feet

Wall

Ladder
(10 feet long)

Floor

Show or explain your work.

Answer _____

34 Every hour, Susanna stuffs 480 envelopes. On average, how many envelopes does Susanna stuff per minute?

Show your work.

Answer _____

35 Rosanna bought 14 compact discs and 12 tapes for her party. She spent a total of $290. Each compact disc cost her $13.

Part A

Write an equation to figure out how much each tape (*t*) cost Rosanna.

Equation _____

Part B

Use the equation you wrote to figure out how much each tape cost Rosanna.

Show your work.

Answer _____

Lidia's Birthday Party

Lidia's 14th birthday party is being held at her house this Saturday. There will be lots of food and activities for all the kids Lidia has invited. Lidia's parents have put her in charge of picking and buying the food, and choosing the activities.

Directions: Numbers 36 through 38 are all about Lidia's birthday party.

36 Lidia's parents gave her $300 to spend on her party. They told her to use 40% of the money on activities, and the rest on food. If the food costs $4.50 per party guest, how many guests can Lidia have at her party?

Show your work.

Number of party guests ____________________

37 One of the activities at Lidia's party will be dodge ball. The party guests are going to be divided into 5 groups: Group 1, Group 2, Group 3, Group 4, and Group 5. Each group has to play every other group one time. How many games of dodge ball will be played?

Show your work.

Answer ________________________________

Session 2

Go to next page

38 Each party invitation included directions to Lidia's house, originating at the corner of Main Street and Maple Drive. That corner was given the coordinates (3,2). Starting there, guests make their first turn at (5,2). Then they turn at (5,5), and then again at (11,5). Lidia's house has the coordinates (11,4).

Part A

On the following grid, show the route the party guests must take to get from the corner of Main Street and Maple Drive to Lidia's house. Plot the points of the route as described above, connect them, and label each point.

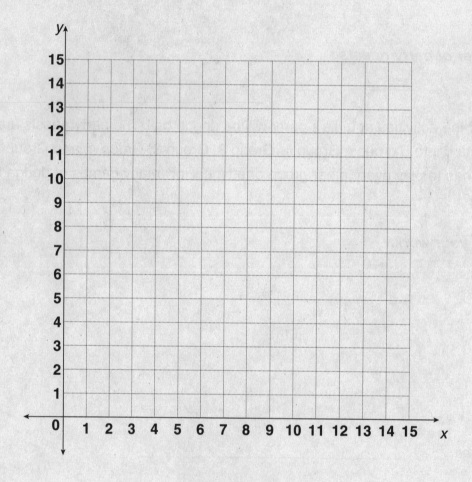

Part B

The distance from the corner of Main Street and Maple Drive to the first turn is 60 yards. How long is the entire route to Lidia's house?

Answer ______________________

Explain how you got your answer.

____

____

____

____

____

____

____

____

Session 2

Go to next page

39 Dennis has to review a 3,000-page medical text for a class he's taking. He's keeping track of how many pages he reviews per day on the following table:

Text Review	
Day	**Number of Pages Reviewed**
Sunday	287
Monday	214
Tuesday	294
Wednesday	256

If Dennis keeps working at about the same rate, ESTIMATE how many *more* days it will take until he finishes reviewing the text.

Show your work.

Answer ______________________

Session 2

Go to next page

40 Use the ruler to help answer this question.

The bottom of a cylinder-shaped can has a radius the length of the line segment below.

_____

Part A

If π is equal to 3.14, and the height of the can is 13 inches, what are the diameter and the volume of the can in cubic inches?

Show your work.

Diameter _____

Volume _____

Part B

Explain how you calculated the diameter and the volume of the can.

Diameter _____

_____

Volume _____

_____

41 On the number line below, graph the following inequality:

$$9 < x \le 17$$

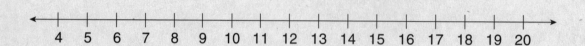

42 Grace and Chang have 36 cups of lemonade to sell. Grace sold $\frac{2}{3}$ of the lemonade, and Chang sold $\frac{1}{4}$ as many cups as Grace sold. How many cups of lemonade did Chang sell?

Show your work.

Answer _____

43 The triangle below, *RST,* is a right triangle.

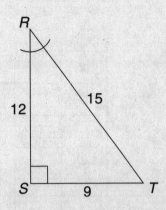

What are the numerical values of the sine and cosine of angle *R*?

Sine R _____

Cosine R _____

44 Fill in the missing numbers in the table below for the ordered pairs for the function $3x + y = 8$.

Part A

x	y
0	
	0
–2	
	–3

Part B

On the following grid, graph the function $3x + y = 8$, labeling the points with the coordinates from the table above:

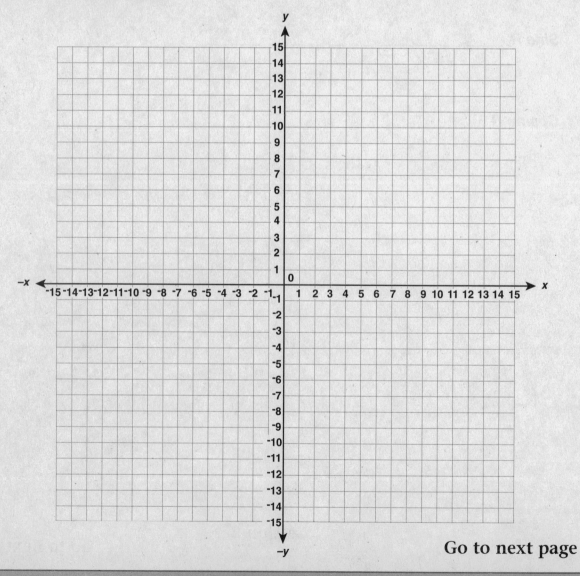

Go to next page

45 A CD player cost the electronics store $120. The store manager marks up the cost of the CD player by 75% to get the price. The sales tax on the CD player is 7%.

Part A

How much would it cost for you to buy the CD player, all totaled, including tax?

Show your work.

Answer _____

Part B

Would you get the same answer if you added the sales tax **before** you marked up the cost of the CD player as you would if you added the sales tax **after** you marked up the cost of the CD player?

Show or explain your work.

Session 2

Stop

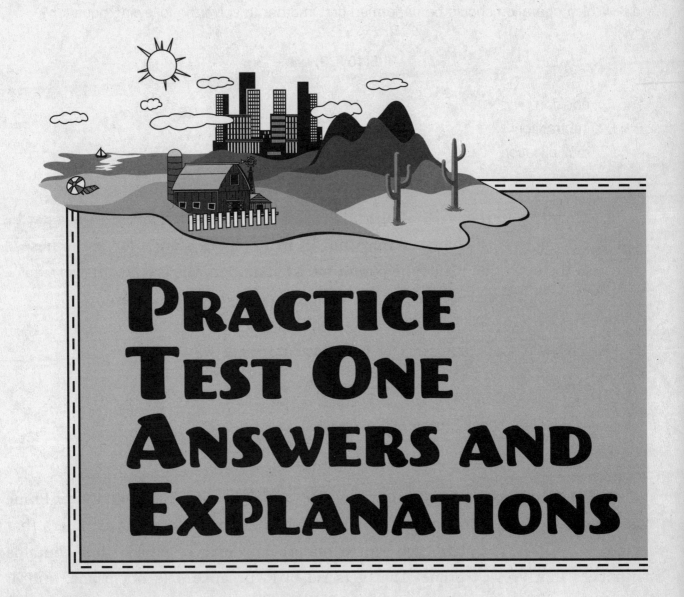

PRACTICE TEST ONE ANSWERS AND EXPLANATIONS

ANSWER KEY FOR MULTIPLE-CHOICE QUESTIONS

1.	A	10.	F	19.	A
2.	G	11.	B	20.	H
3.	D	12.	G	21.	B
4.	H	13.	B	22.	H
5.	C	14.	H	23.	A
6.	J	15.	A	24.	H
7.	D	16.	F	25.	B
8.	G	17.	C	26.	G
9.	C	18.	H	27.	B

SESSION I, PART I

1   Which operation should be performed first in order to solve the following equation?

$$5(10 + 2) - 7 \div 3 =$$

A   addition
B   subtraction
C   multiplication
D   division

A According to the order of operations, whatever's inside parentheses must be taken care of first. (Remember that the "P" in PEMDAS stands for parentheses.) Because the operation inside the parentheses is addition, the correct answer choice is A. Notice that you don't have to solve the equation to get the answer.

2   Which one of these represents the prime factorization of 36?

F 4×3^2
G $2^2 \times 3^2$
H $2^3 \times 3^2$
J 6^2

G This is a good opportunity to use the Process of Elimination. Prime factorization means that a value must be broken down into a set of numbers that must be prime. Therefore, you can eliminate any answer choice that contains numbers that are *not* prime. That gets rid of F, because 4 is not prime, and J, because 6 is not prime. Now you're left with two choices, G and H. Now check out the values of choice G. 2^2 is 4, and 3^2 is 9. $4 \times 9 = 36$. That's it!

3   A wooden ruler is marked to indicate the lengths $\frac{1}{2}$, $\frac{2}{3}$, $\frac{1}{3}$, and $\frac{1}{5}$ of an inch. Which of these shows the order of these marks, from the largest length to the smallest?

A $\frac{1}{2}$, $\frac{1}{3}$, $\frac{1}{5}$, $\frac{2}{3}$

B $\frac{1}{2}$, $\frac{1}{3}$, $\frac{2}{3}$, $\frac{1}{5}$

C $\frac{1}{5}$, $\frac{2}{3}$, $\frac{1}{3}$, $\frac{1}{2}$

D $\frac{2}{3}$, $\frac{1}{2}$, $\frac{1}{3}$, $\frac{1}{5}$

D One way to solve this problem is to compare the fractions in pairs. Or you can give each fraction a common denominator; 30 would be the smallest value you can use. But you might also realize that if you're comparing a few fractions that all have the same numerator (in this case, 1), the bigger the denominator, the smaller the value of the fraction. That means that $\frac{1}{5}$ has the smallest value. Next comes $\frac{1}{3}$, then $\frac{1}{2}$. Finally, $\frac{2}{3}$ is the biggest. Now, the question asks for the order from largest to smallest, so the order is $\frac{2}{3}$, $\frac{1}{2}$, $\frac{1}{3}$, $\frac{1}{5}$. Another way to solve this problem would be to change the fractions to decimals. That means $\frac{1}{2}$ = 0.5, $\frac{1}{3}$ = $0.3\overline{3}$, $\frac{2}{3}$ = $0.6\overline{6}$, and $\frac{1}{5}$ = 0.2. Now put the values in order from largest to smallest: $0.6\overline{6}$, 0.5, $0.3\overline{3}$, and 0.2. Now, just match the decimals to the fractions.

4   To make the following equation true, which number goes in the box?

$$5^{\square} = 625$$

F   2
G   3
H   4
J   5

H This problem offers you a great opportunity to use the answer choices for help. Start with F, and put 2 in the box. Does $5^2 = 625$? No, $5^2 = 25$. 5^3 is $5 \times 5 \times 5$, which is 125. That's still too small. 5^4 is $5 \times 5 \times 5 \times 5$, which is 625. That's it.

5   Kathy and Janine sold some brownies on Saturday. Kathy sold $\frac{1}{4}$ of the brownies, and Janine sold 40% of the brownies. What percentage of the brownies did Kathy and Janine sell altogether?

    A   44%

    B   50%

    C   65%

    D   75%

C Because the question asks for a percentage, you should convert the fraction $\frac{1}{4}$ to a percentage. The 40% is already a percentage. $\frac{1}{4}$ is equal to 25%. Therefore, the total percentage is 25% + 40%, which is 65%.

6   The table at a board meeting is shaped like this.

The table is known to have a perimeter of 95 inches. How long is the longest side of the table if the side opposite it is 32 inches long, and the other sides are each 12 inches?

    F   12

    G   24

    H   32

    J   39

J Perimeter is the sum of the lengths of all the sides. The question provides the perimeter of the table and the lengths of three out of its four sides. Now you know that 95 = 32 + 12 + 12 + longest side. Simplify that: 95 = 56 + longest side. 95 – 56 = 39, so the longest side is 39.

7   Nikolas ate $\frac{4}{5}$ of his sandwich. What percentage of his sandwich did Nikolas eat?

A   4.5%

B   20%

C   45%

D   80%

D Because this question asks for a percentage, you have to convert $\frac{4}{5}$ to a percentage: $\frac{4}{5} = \frac{x}{100}$. Cross multiply and you will get $5x = 400$. To isolate, divide both sides by 5. You will get $x = 80$.

8   Use your ruler to solve this problem.

A rectangular pool is shown below.

scale: 1 inch = 1 yard

Based on the scale, what is the perimeter of the pool?

F   9 yards

G   10 yards

H   10.5 yards

J   11 yards

G Use the ruler to measure the sides of the picture, and you should get a length of 3 inches and a width of 2 inches. Perimeter is the sum of the lengths of the figure, which in this case is 3 + 2 + 3 + 2, or 10. Because the scale tells us that 1 inch equals 1 yard, the answer is 10 yards.

---

9   In the diagram below, trace over the line segments *AB* and *CD*. Which angle is formed by the intersection of line segments *CD* and *AB*?

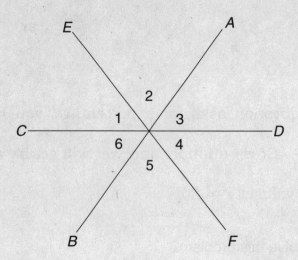

A   1

B   2

C   3

D   4

C The angles formed at the intersection of line segments *CD* and *AB* are angles 3 and 6. Because 6 is not among the answer choices, the answer must be 3.

10   Juan and Simone are buying fruit, and each is looking at an apple. Juan says, "I'll have 5 pieces of fruit if I buy this apple and two more." Simone says, "And I'll have 4 pieces of fruit if I buy this apple and one more." Which of these is correct?

F   Juan has 2 pieces of fruit.

G   Simone has 3 pieces of fruit.

H   Juan has 3 pieces of fruit.

J   Simone has 4 pieces of fruit.

F First start with Juan's situation. He says he'll have 5 pieces of fruit if he buys the 1 apple and 2 more. The 1 plus the 2 more is 3. So, if he'll have 5 pieces of fruit if he adds 3 to what he already has, he must already have 2. Hey, check the answers—that's choice F. You don't have to worry about Simone because you've already found a correct answer. It pays to solve word problems one step at a time, and to use the answer choices.

11 Leon is figuring out how much a long-distance call will cost him. He gets charged $0.50 for the first minute he speaks, and $0.35 for each additional minute. How long can he speak if the call costs him $8.20?

   A   22 minutes
   B   23 minutes
   C   25 minutes
   D   26 minutes

B The call costs Leon $8.20. Subtract away that 50 cents for the first minute right away (but don't forget about it, because it counts as a minute!). $8.20 – $0.50 = $7.70. Now, take that $7.70 and divide it by the cost of each additional minute, 35 cents, to get the number of remaining minutes. $7.70 ÷ 0.35 = 22 minutes. But wait, don't pick A. You first have to add the one minute from that "50 cents for the first minute," so the answer is 23 minutes.

12 Everett is building a fence around his yard that will look like the figure below.

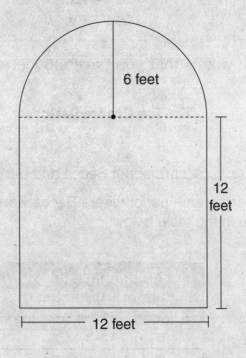

How much material does he need for the fence?

(use π = 3.14)

   F   50.68 feet
   G   54.84 feet
   H   66.84 feet
   J   73.68 feet

G This question requires you to find the perimeter of Everett's yard. The three straight sides are all 12 feet, so that adds up to 36 feet. Now for the curved part. This is a semicircle, because the radius is shown to be 6 in two places. If it were a whole circle, you would find the circumference with the formula $C = 2\pi r$. So for a semicircle, it's $C = \dfrac{2\pi r}{2}$. The radius is 6, so solve for C.

$$C = \frac{2\pi r}{2}$$

$$C = \frac{(2)(3.14)(6)}{2}$$

$$C = 37.68 \div 2$$

$$C = 18.84 \text{ feet}$$

Okay, now add up the values that you have: 36 feet + 18.84 feet = 54.84 feet.

The Ice Cream Parlor

Directions: Numbers 13 through 16 are about Becca and her job at the ice cream parlor.

13 On Wednesdays at the ice cream parlor where Becca works, there's a special deal on free toppings, as shown in the table below.

Free Topping Wednesday	
Number of Scoops	Number of Free Toppings
1	2
2	5
3	7
4	10
5	12

If Ms. Meyers receives 17 free toppings with the ice cream she buys on Wednesday, and the pattern continues, how many scoops of ice cream did she buy?

A 6

B 7

C 8

D 9

B Look for the pattern in the free toppings. The difference between 2 and 5 is 3. The difference between 5 and 7 is 2. The difference between 7 and 10 is 3. The difference 10 and 12 is 2. So, the pattern is + 3, + 2, + 3, + 2. That means that the next number of free toppings (which would correspond with 6 scoops) would be 12 + 3, or 15, and the following number (which would correspond with 7 scoops) would be 15 + 2, or 17. That's what we're looking for.

14 Becca works at the ice cream parlor Monday through Friday, four hours each day, and gets paid $6.50 an hour. How much money has she made after three weeks?

F $130

G $162.50

H $390

J $546

H Becca works 5 days a week for 4 hours each day. That's 5 × 4, or 20 hours a week. She gets paid $6.50 an hour, so how much does she make in a week? $6.50 × 20 hours, or $130. Wait, don't pick F. The question asks for how much she makes after *three* weeks. That's $130 × 3, or $390, answer choice H. It pays to read the question carefully!

15 Becca is keeping track of the number of ice-cream cones of each flavor she sells this week in the following table:

Weekly Cone Sales	
Flavor	Number of cones sold
chocolate	10
vanilla	12
strawberry	4
coffee	5
pistachio	0
rocky road	6
fudge almond	8
pralines and cream	5

What is the **mean** number of cones of each flavor Becca sold this week?

A 6.25

B 6.5

C 7

D 7.5

A To find the mean (which is the same thing as the average), add up the total, and divide by the number of things. So, 10 + 12 + 4 + 5 + 0 + 6 + 8 + 5 = 50. Now, divide 50 by the number of flavors, which is 8 (don't forget to include pistachio, even though 0 cones were sold). 50 ÷ 8 = 6.25.

16 Becca's friend Joey comes into the ice cream parlor every Friday to buy a pint of ice cream. This Friday, there are 14 pints of rocky road and 12 pints of fudge almond in the freezer, and no other flavors. If he chooses a pint at random, what is the probability that Joey will buy a pint of rocky road?

F $\dfrac{7}{13}$

G $\dfrac{6}{13}$

H $\dfrac{6}{7}$

J $\dfrac{7}{6}$

F There is a total of 26 pints of ice cream in the freezer, and 14 of them are rocky road. The probability of Joey choosing rocky road at random is $\frac{14}{26}$, which reduces to $\frac{7}{13}$. You could have eliminated choice J right away; probability can never be greater that 1.

17 In right triangle *EFG* below, which of the following is a true statement?

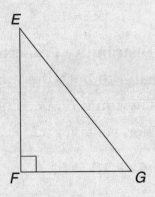

A   The sum of angles *E* and *F* is 90 degrees.
B   The sum of angles *F* and *G* is 180 degrees.
C   The sum of angles *E* and *G* is 90 degrees.
D   The sum of angles *E* and *F* is 180 degrees.

C The sum of the angles in any triangle is 180 degrees. Angle F is a right angle, which measures 90 degrees. That means the other two angles, E and G, must add up to the remaining 90 degrees.

18 Which value for *x* will make the statement below true?

$$2 + 4(x + 2) = 14$$

F   −1
G   0
H   1
J   2

H Your job is to solve for *x*. First, subtract 2 from both sides.

$$
\begin{array}{r}
2 + 4(x + 2) = 14 \\
\underline{-2 \qquad\qquad\quad -2} \\
4(x + 2) = 12
\end{array}
$$

Then divide both sides by 4.

$$\frac{4(x+2)}{4} = \frac{12}{4}$$

$$x + 2 = 3$$

Then subtract 2 from both sides.

$$x + 2 = 3$$
$$\underline{-\ 2 - 2}$$
$$x\ =\ 1$$

You also could have used the Distributive Property on the original equation, $2 + 4(x + 2) = 14$. Distribute that 4 to both the numbers in the parentheses by multiplying $4 \times x$ and 4×2. The equation becomes $2 + 4x + 8 = 14$. That can be simplified into $4x + 10 = 14$. Then solve for x.

$$4x + 10 = 14$$
$$\underline{-\ 10 - 10}$$
$$4x = 4$$

$$\frac{4x}{4} = \frac{4}{4}$$

$$x\ =\ 1$$

19

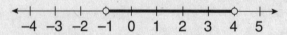

Which of these inequalities is represented on the number line above?

A $-1 < x < 4$

B $-1 \leq x \leq 4$

C $-1 \geq x > 4$

D $-1 < x \leq 4$

A The empty circles on the number line represent either less than (<) or greater than (>). The bold line connects –1 to 4. So, the number line represents all of the numbers between (but not including) –1 and 4. That's A. But just by noticing that both circles are empty on the number line, you could have eliminated choices B, C, and D, because they all contain either ≤ or ≥ (which are represented by solid circles).

20 During her seven-hour shift at the video store, Keesha has to rewind videotapes. She must keep track of the number of videotapes she rewinds each hour.

Videotape Rewinding	
Hour	Numbers of Tapes
1	42
2	35
3	48
4	32
5	29
6	51
7	41

What is the *median* number of tapes Keesha rewound during her seven-hour shift?

F 32

G 39

H 41

J 48

H The median is the middle number when the numbers are in order, so start by putting the numbers in order from least to greatest: 29, 32, 35, 41, 42, 48, and 51. Now, which number is directly in the middle? It's 41, answer choice H.

21 Ling has a small six-sided block. Two of the faces are painted red, one is painted blue, one is painted yellow, one is painted green, and one is painted orange. If Ling throws the block up in the air, what is the probability that it will land with the green side facing up?

A $\frac{1}{25}$

B $\frac{1}{6}$

C $\frac{1}{5}$

D $\frac{1}{4}$

B The block has six faces. Only one of them is painted green. You can find the fraction of probability by using the formula $P = \dfrac{\text{number of ways event can occur}}{\text{number of possible outcomes}}$.

So the probability that the block would land with the green face up is $\dfrac{1}{6}$.

22 A recipe calls for 2 eggs for every 3 cups of flour. Which of these proportions shows the way to find how many eggs, *e*, are needed for a recipe calling for 27 cups of flour?

F $\dfrac{e}{27} = \dfrac{3}{2}$

G $\dfrac{27}{e} = \dfrac{2}{3}$

H $\dfrac{e}{27} = \dfrac{2}{3}$

J $\dfrac{27}{3} = \dfrac{2}{e}$

H The key to setting up a proportion is being careful to put like terms in the same places.

$$\frac{2 \text{ eggs}}{3 \text{ cups of flour}} = \frac{e \text{ eggs}}{27 \text{ cups of flour}}$$

That's H. Notice that you're not asked to actually solve for *e*. Don't do more work than you have to.

23 Which of the following shows a pair of figures that are *similar* but not *congruent?*

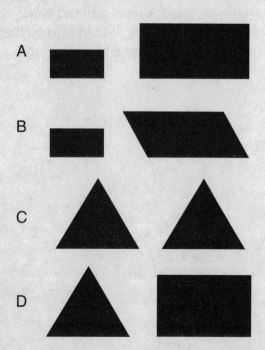

A In this case, you must go through the shapes in the answer choices and evaluate them. Notice that the question asks about the definitions of **similar** and **congruent.** Similar figures have the same shape. That eliminates B and D. But similar figures can differ in size, whereas congruent figures have exactly the same shape and size. That means that the figures in choice C are congruent, and C can be eliminated. That leaves A as the correct answer choice.

24 Maria buys a bag of jelly beans to share with her friend Pierre. Each bag contains two black jelly beans, which Maria always gives to Pierre. If m equals the number of jelly beans in a bag, and p equals the number after Maria gives Pierre the two black jelly beans, which of the equations below would you use to figure out how many jelly beans are left in the bag after Maria gives Pierre the black ones?

F $p = 2 + m$
G $p - m = 2$
H $p = m - 2$
J $m = p - 2$

H The question says that m is the number of jelly beans in the bag to begin with. You're also told that Maria gives two of those jelly beans to Pierre. Finally, the question says that p represents the number of beans in the bag *after* those two beans are given to Pierre. So, the number of beans in the bag, minus 2, equals the number after the two are given away. In other words, $m - 2 = p$. That's the same as H.

25 Bobbi has a job making sales phone calls. Besides getting $5.25 an hour, she also receives an additional $0.30 for each phone call she makes. On Tuesday, Bobbi earned $18.45 and made an average of three phone calls per hour. How many hours did she work on Tuesday?

A   2 hours
B   3 hours
C   4 hours
D   4.5 hours

B This is a good opportunity to use the answer choices. Start with A, and say that Bobbi worked 2 hours on Tuesday. See if that checks out with all of the other information in the question. If it does, you're done. If it doesn't, you can eliminate choice A. Suppose Bobbi worked 2 hours. She gets $5.25 per hour: ($5.25)(2) = $10.50. She also made 3 calls per hour, so if she worked 2 hours, that's 6 phone calls. She gets $0.30 for each phone call: ($0.30)(6) = $1.80. Now you have to see if $10.50 + $1.80 = $18.45, the amount she earned on Tuesday. Does $12.30 = $18.45? No. Now you know the correct answer choice is not A. Try B and say that Bobbi worked 3 hours. She gets $5.25 per hour: ($5.25)(3) = $15.75. She also made 3 calls per hour, so if she worked 3 hours, that's 9 phone calls. She gets $0.30 for each phone call: ($0.30)(9) = $2.70. All you have to do now is see if $15.75 + $2.70 = $18.45, the amount she earned on Tuesday. Does $18.45 = $18.45? Yes. The correct answer choice is B.

26 If Kevin can guess how many beans are in a jar, he'll win a prize. He knows the following: There are only red, orange, yellow, and green beans in the jar. The jar contains twice as many orange beans as red beans, three times as many green beans as yellow beans, and half as many yellow beans as red beans. If there are 6 red beans, how many beans are in the jar?

F 20

G 30

H 35

J 40

G Turn those sentences into equations: There are twice as many orange beans as red beans, so $O = 2R$. There are three times as many green beans as yellow beans, so $G = 3Y$. There are half as many yellow beans as red beans, so $Y = \dfrac{R}{2}$. The good news is that the question tells you what R is—it's 6. Now start solving by substituting what you already know. Because R is 6, substitute the 6 into the equation $Y = \dfrac{R}{2}$ to solve for Y.

$$Y = \frac{R}{2}$$

$$Y = \frac{6}{2}$$

$$Y = 3$$

So there are 3 yellow beans. Because $Y = 3$, substitute 3 into $G = 3Y$ to solve for G.

$$G = 3Y$$

$$G = 3(3)$$

$$G = 9$$

So there are 9 green beans. Because R is 6, substitute the 6 into the equation $O = 2R$ to solve for O.

$$O = 2R$$
$$O = 2(6)$$
$$O = 12$$

So there are 12 orange beans.

Now, add up the beans: 6 (red) + 3 (yellow) + 9 (green) + 12 (orange) = 30 beans.

27 On the grid below, one triangle is shown. Two coordinates for another triangle are also shown.

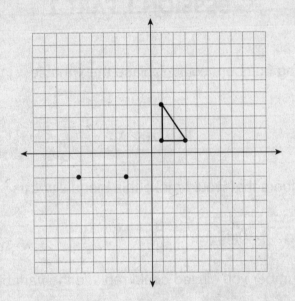

Which of these sets of coordinates will complete a second triangle that is similar to the triangle shown?

A (−2,−6)

B (−2,−8)

C (−2,−9)

D (−2,−7)

B First, count the boxes to get the line lengths of the shown triangle: It has a base of 2 and a height of 3. Now connect the dots on that other line. Count the boxes to find its length: 4. That line forms a part of a triangle that has to be similar to the first triangle. The 4 has a relationship to the 2: It's twice as much. So, the height of the second triangle should be twice 3, the height of the first triangle. That would be 6.

Now, take a look at the answer choices. They all have something in common, don't they? They all have an *x*-coordinate of –2. Because the missing coordinate should form a line that is 6 boxes long, count 6 boxes from –2. You wind up at either (–2,–8) or (–2,4). Though (–2,4) isn't among the choices, (–2,–8) is; therefore, the answer is B.

SESSION I, PART 2

28 Which of the following numbers could replace the variable *x* in the inequality below?

$$0.67 > x > \frac{2}{5}$$

Circle all of the numbers that would make this inequality true.

$$\frac{1}{3} \qquad 0.42 \qquad \frac{6}{8} \qquad \frac{1}{2} \qquad \frac{3}{2}$$

Explain why each number you circled could replace the variable *x*.

Explanation

This question would be scored with the two-point scale. For the first part of the question, you need to circle all of the numbers that could replace *x* in the inequality $0.67 > x > \frac{2}{5}$. One way to start is to convert $\frac{2}{5}$ to 0.4. Now you know you're looking for any values between 0.67 and 0.4. That eliminates $\frac{1}{3}$ (or $.3\overline{3}$). 0.42 fits the range, so that number should be circled. $\frac{6}{8}$ is the same as $\frac{3}{4}$, which is the same as 0.75, which is too big to fit in the range. $\frac{1}{2}$ is the same as 0.5, which is between 0.4 and 0.67, so you should circle that number. Finally, $\frac{3}{2}$ is bigger than 1, so that can't work. So, you should have circled only 0.42 and $\frac{1}{2}$. If you find fractions easier to work with than decimals, you could have converted everything to fractions and compared them that way. Either way 0.42 and $\frac{1}{2}$ are correct.

For the second part of the question, you have to explain how you got your answers. Explain in as much detail as you can; remember, when in doubt, write it out (WIDWIO). You can either explain it in sentences or mathematically, such as "I circled 0.42 and $\frac{1}{2}$ because they are both between 0.67 and $\frac{2}{5}$, which is the same as 0.4" and "$\frac{1}{3} < \frac{2}{5}$, so this doesn't work," etc. Your explanation must include something about how the circled numbers are between 0.67 and $\frac{2}{5}$.

29

$$m \div k = k$$
$$n + m = j$$
$$j \times n = j$$
$$n + n = k$$

In the three equations above, variables $j, k, m,$ and n each represent a different whole number. If $k = 2$, find the value for each of the remaining variables. Show all of your work.

Answers

$j =$ _____

$m =$ _____

$n =$ _____

Explanation

This question would be scored with the two-point scale. Be sure to show your work in the space provided. Explain in as much detail as you can. The problem states that $k = 2$, so look for an equation that involves a k. How about $n + n = k$? Make that $n + n = 2$, which means that $n = 1$. Fill that in where it says "Answers" so you don't forget. $n = 1$ also makes sense when you look at $j \times n = j$; even though you don't know what j is yet, you know that anything times 1 is itself. Now, there's another equation with k: $m \div k = k$. In other words, m divided by 2 equals 2. That means that $2 \times 2 = m$, so $m = 4$. Fill that in where it says "Answers" so you don't forget. Now you know what n and m are, so do $n + m = j$. That's $1 + 4 = j$, so j is 5. Fill that in where it says "Answers" so you don't forget. So $j = 5$, $m = 4$, and $n = 1$.

30 The members of the marching band are selling T-shirts and sweatshirts to raise money for new uniforms. The following table shows the number of T-shirts and sweatshirts the band sells over a four-day period:

T-Shirt and Sweatshirt Sales		
Day	Number of T-Shirts	Number of Sweatshirts
Monday	42	21
Tuesday	24	28
Wednesday	30	32
Thursday	38	25

Part A

Construct a double line graph of the information from the table on the following grid, being sure to

- title the graph and label the axes
- be accurate
- make a key
- be consistent with the scale

T-Shirt and Sweatshirt Sales

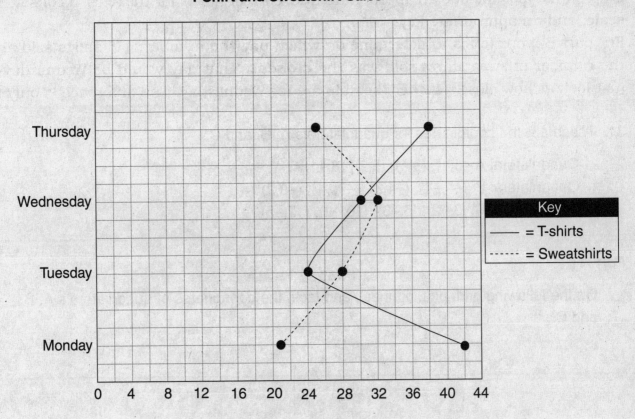

Part B

Using your double line graph, determine on which day the number of T-shirts sold and the number of sweatshirts sold was the *closest*.

Day _____

Explanation

This question would be scored with the three-point scale. In part A, your job is to construct a graph based on the information you're given about the T-shirt and sweatshirt sales. Here is one example of what it could look like. (It can also be set up so that the days of the week are on the *x*-axis.)

Besides the actual information about the sales of T-shirts and sweatshirts, your graph must also include an appropriate title, correct labels for the axes, a consistent scale, and an appropriate key.

For part B, your job is to determine on which day the number of T-shirts sold and the number of sweatshirts sold was the **closest.** That day would be Wednesday— just look at how close together the dots are for Wednesday. The difference is only 2.

31 These are the coordinates for quadrilaterals A, B, and C.

Quadrilateral A: (5,4), (8,4), (5,9), (8,11)
Quadrilateral B: (−4,−1), (−4,2), (−9,−1), (−9,2)
Quadrilateral C: (3,−2), (5,−5), (10,−2), (12,−5)

Part A

On the following grid, plot, connect, and label the coordinates of quadrilaterals A, B, and C.

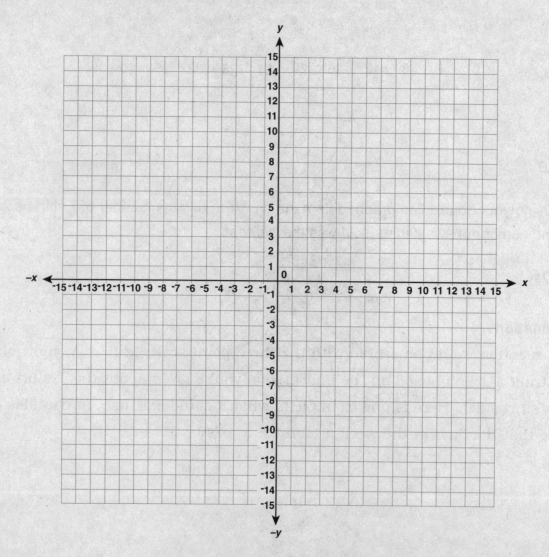

Part B

Name and describe what type(s) of quadrilaterals each one is, and explain in words those qualities that helped you identify them.

Quadrilateral A is a _____

Quadrilateral B is a _____

Quadrilateral C is a _____

Explanation

This question would be scored with the three-point scale. For part A, your job is to graph these quadrilaterals. That should look like this (note that each point is labeled).

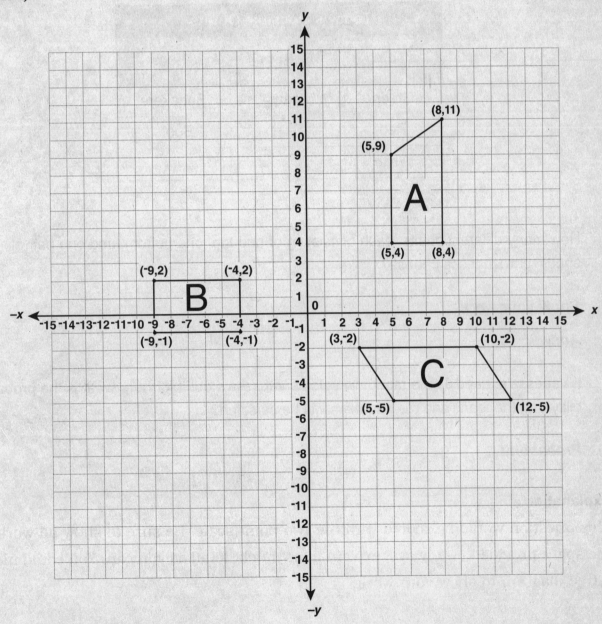

For part B, you need to identify each quadrilateral and explain how you did so. Explain in as much detail as you can. Quadrilateral A is a trapezoid because it has exactly one pair of opposite sides that are parallel. Quadrilateral B is a rectangle because it has two pairs of opposite sides that are parallel and four right angles. Quadrilateral B is also considered a parallelogram because it has two pairs of opposite sides that are parallel. Quadrilateral C is a parallelogram because it has two pairs of opposite sides that are parallel and the opposite angles are equal.

32 Vanessa is deciding which outfit to wear to school today. Here is what she has to choose from.

Today's Possible Outfits		
Top	**Bottom**	**Shoes**
Green	Jeans	Sneakers
Red	Skirt	Boots
White		

How many different outfits, each consisting of one top, one bottom, and one pair of shoes, can Vanessa choose?

Show your work.

Answer _____

If Vanessa's friend Jamie randomly selects Vanessa's outfit for her, what is the probability that the jeans and the sneakers will be selected?

Probability _____

Explanation

This question would be scored with the two-point scale. Be sure to show all work in the space provided. As always, explain in as much detail as you can. You could make a tree diagram to show your work.

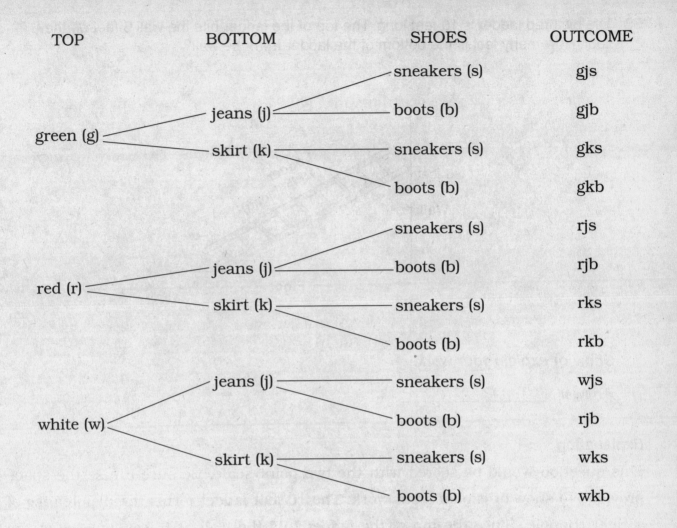

| | TOP | BOTTOM | SHOES | OUTCOME |

TOP — BOTTOM — SHOES — OUTCOME

green (g)
- jeans (j)
  - sneakers (s) — gjs
  - boots (b) — gjb
- skirt (k)
  - sneakers (s) — gks
  - boots (b) — gkb

red (r)
- jeans (j)
  - sneakers (s) — rjs
  - boots (b) — rjb
- skirt (k)
  - sneakers (s) — rks
  - boots (b) — rkb

white (w)
- jeans (j)
  - sneakers (s) — wjs
  - boots (b) — rjb
- skirt (k)
  - sneakers (s) — wks
  - boots (b) — wkb

There are 12 different outfits, or outcomes.

You also could have used the Counting Principle: If event M can occur in m ways and is followed by event N that can occur in n ways, then the event M followed by the event N can occur in $m \times n$ ways. In the case of the outfits, that's the number of choices for tops, which is 3, times the number of choices for bottoms, which is 2, times the number of choices for shoes, which is 2. So, $3 \times 2 \times 2 = 12$.

For the second part of the question, you need to figure out the probability that the jeans and the sneakers were randomly chosen for Vanessa by Jamie. The tree diagram from the first part of the question showed that there were 12 possible outfits. The question is, how many of them include both the sneakers and the jeans? There's gjs (the jeans and sneakers with the green top), rjs (the jeans and sneakers with the red top), and wjs (the jeans and sneakers with the white top). That's 3 out of a possible 12 outfits. The answer is $\frac{3}{12}$, which can be reduced to $\frac{1}{4}$ or 0.25.

33 The pictured ladder is 10 feet long. The top of the ladder hits the wall 6 feet off the floor. How many feet is the bottom of the ladder from the wall?

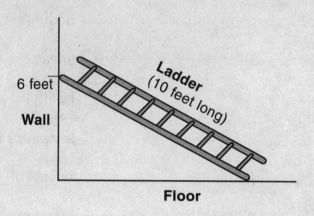

Show or explain your work.

Answer _____

Explanation

This question would be scored with the two-point scale. Be sure to use the space provided to show or explain your work. The 10-foot ladder forms the hypotenuse of a right triangle. Since the top of the ladder hits the wall at 6 feet, one leg of the triangle is 6 feet. Use the Pythagorean theorem to find the length of the missing side, which is also the number of feet the bottom of the ladder is from the wall.

$$c^2 = a^2 + b^2$$
$$10^2 = 6^2 + b^2$$
$$100 = 36 + b^2$$
$$64 = b^2$$
$$\sqrt{64} = b$$
$$8 = b$$

The answer is 8 feet (don't forget the word "feet").

You could also have solved this by recognizing that in a right triangle with a leg of 6 and a hypotenuse of 10, the other leg would have to be 8 because the 6-8-10 right triangle is a multiple of the primitive Pythagorean triple, 3-4-5. That would also be a valid explanation.

34 Every hour, Susanna stuffs 480 envelopes. On average, how many envelopes does Susanna stuff per minute?

Show your work.

Answer _____

Explanation

This question would be scored with the two-point scale. Be sure to show all work in the space provided. Explain in as much detail as you can. This is a proportion question. Susanna stuffs 480 envelopes every hour. The question asks how many she stuffs per minute, so set up the proportion as envelopes per minute. In other words, instead of 480 envelopes per 1 hour, make it 480 envelopes per 60 minutes.

$$\frac{480}{60} = \frac{x}{1}$$

Now cross multiply, and you get $60x = 480$. Divide both sides by 60, and you get 8. The answer is 8 envelopes per minute (don't forget to add "envelopes per minute").

35 Rosanna bought 14 compact discs and 12 tapes for her party. She spent a total of $290. Each compact disc cost her $13.

Part A

Write an equation to figure out how much each tape (*t*) cost Rosanna.

Equation _____

Part B

Use the equation you wrote to figure out how much each tape cost Rosanna.

Show your work.

Answer _____

Explanation

This question would be scored with the two-point scale. For part A, you have to write an equation to figure out how much each tape cost Rosanna. The question says that she bought 14 compact discs and 12 tapes for her party, she spent a total of $290, and each compact disc cost her $13. So $290 = (the 14 compact discs × $13) + (the 12 tapes × *t*). In other words,

$$\$290 = (14 \times \$13) + (12 \times t).$$

Your equation can look different and still be correct. You could have the $290 on the right side of the equation, or you could have the (12 × *t*) to the left of the (14 × $13), or you could use different-looking brackets. As long as your equation is mathematically sound, you're fine.

For part B, you get to solve for *t*. Be sure to show all work in the space provided, and be consistent with those $ signs. As usual, explain in as much detail as you can.

$$\$290 = (14 \times \$13) + (12 \times t)$$
$$\$290 = \$182 + 12t$$
$$\underline{-\$182 \quad -\$182}$$
$$\$108 = 12t$$
$$\frac{\$108}{12} = \frac{12t}{12}$$
$$\$9 = t$$

So each tape cost $9.

Lidia's Birthday Party

Lidia's 14th birthday party is being held at her house this Saturday. There will be lots of food and activities for all the kids Lidia has invited. Lidia's parents have put her in charge of picking and buying the food, and choosing the activities.

Directions: Numbers 36 through 38 are all about Lidia's birthday party.

36 Lidia's parents gave her $300 to spend on her party. They told her to use 40% of the money on activities, and the rest on food. If the food costs $4.50 per party guest, how many guests can Lidia have at her party?

Show your work.

Number of party guests _____

Explanation

This question would be scored with the two-point scale. Be sure to show all work in the space provided. Explain in as much detail as you can; remember, when in doubt, write it out (WIDWIO). Lidia has $300, and she is to use 40% of the money on activities, and the rest on food. Stop there. If she uses 40% for activities, what percent does she use for food? 60%. Now, what is 60% of $300? Remember, you can use your calculator on this section (but be sure to write down what you've calculated!), so just punch in 0.60×300, and you will get 180. So Lidia spends $180 on food. Now, you're told that the food cost $4.50 per guest, so to find out how many guests she can have, divide $180 by $4.50. You will get 40. The answer is 40 guests.

37 One of the activities at Lidia's party will be dodge ball. The party guests are going to be divided into 5 groups: Group 1, Group 2, Group 3, Group 4, and Group 5. Each group has to play every other group one time. How many games of dodge ball will be played?

Show your work.

Answer _____

Explanation

This question would be scored with the two-point scale. Be sure to show all work in the space provided. Explain in as much detail as you can; remember, when in doubt, write it out (WIDWIO). You could make a tree diagram.

TEAM 1	TEAM 2	OUTCOME
	2	12
	3	13
1	4	14
	5	15
	3	23
2	4	24
	5	25
	4	34
3	5	35
4	5	45

The answer is 10, because there are 10 outcomes.

This is a combination, so you also could have used the formula $_nC_r = \dfrac{n!}{(n-r)!\,(r!)}$.

There are 5 teams playing 2 at a time. Plug that into the formula.

$$_5C_2 = \frac{5!}{(5-2)!\,(2!)}$$

$$\frac{5 \times 4 \times 3 \times 2 \times 1}{(3 \times 2 \times 1)(2 \times 1)}$$

Don't forget to reduce.

$$\frac{5 \times 4}{2 \times 1} = \frac{20}{2} = 10$$

The answer is 10.

38 Each party invitation included directions to Lidia's house, originating at the corner of Main Street and Maple Drive. That corner was given the coordinates (3,2). Starting there, guests make their first turn at (5,2). Then they turn at (5,5), and then again at (11,5). Lidia's house has the coordinates (11,4).

Part A

On the following grid, show the route the party guests must take to get from the corner of Main Street and Maple Drive to Lidia's house. Plot the points of the route as described above, connect them, and label each point.

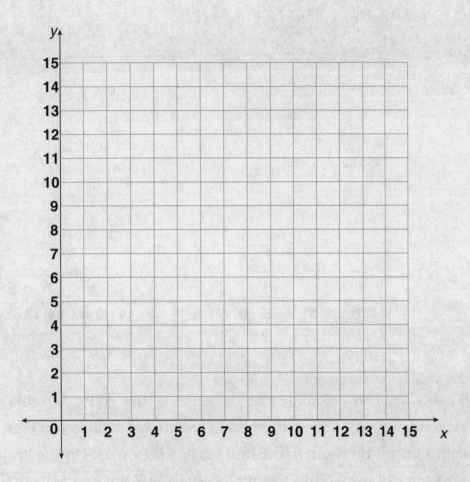

Part B

The distance from the corner of Main Street and Maple Drive to the first turn is 60 yards. How long is the entire route to Lidia's house?

Answer _____

Explain how you got your answer.

Explanation

This question would be scored with the three-point scale. For part A, your job is to graph the coordinates of each stop described in the directions. This is what it should look like (note that each point is labeled).

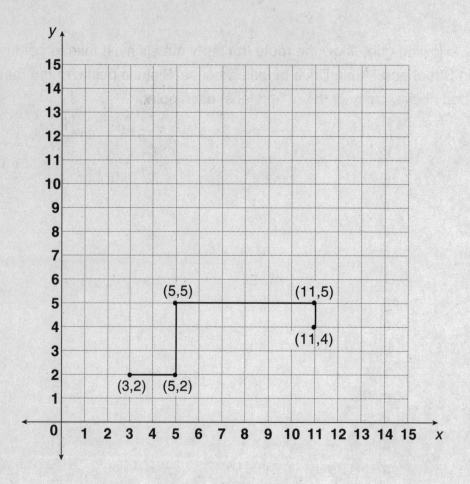

For part B, you need to calculate the distance of the route to Lidia's house, and explain how you got your answer. Explain in as much detail as you can. The distance from the beginning of the route to the first turn is 60 yards. On the graph, that same part of the route is 2 boxes. That means the scale is 2 boxes = 60 yards, or each box equals 30 yards. Now count how many boxes make up the entire route from start to finish. It's 12 boxes. If 1 box = 30 yards, 12 boxes = 12 × 30 or 360 yards. The answer is 360 yards.

You also could have figured that since the scale is 2 boxes = 60 yards, and there are 12 boxes, the proportion is $\frac{2}{60} = \frac{12}{x}$. Cross multiply and you get $2x = 720$. The answer is still 360 yards.

39 Dennis has to review a 3,000-page medical text for a class he's taking. He's keeping track of how many pages he reviews per day on the following table:

Text Review	
Day	Number of Pages Reviewed
Sunday	287
Monday	214
Tuesday	294
Wednesday	256

If Dennis keeps working at about the same rate, ESTIMATE how many *more* days it will take until he finishes reviewing the text.

Show your work.

Answer _____

Explanation

This question would be scored with the three-point scale. Be sure to show all work in the space given. Explain in as much detail as you can. Note that you're being asked to ESTIMATE. You could start by rounding to an approximate daily page amount, say, 250. Dennis has 3,000 pages to review, so $3,000 \div 250$ is 12 days. He has already reviewed for 4 days, so that leaves 8 days.

Here's another way: You could start by rounding to an approximate daily page amount, say, 250. He has read for four days, so that's about 1,000 pages. He has 2,000 more pages to read, so that would take him 4×2 or 8 more days. The answer is still 8 days.

40 Use the ruler to help answer this question.

The bottom of a cylinder-shaped can has a radius the length of the line segment below.

_____

Part A

If π is equal to 3.14, and the height of the can is 13 inches, what are the diameter and the volume of the can in cubic inches?

Show your work.

Diameter _____

Volume _____

Part B

Explain how you calculated the diameter and the volume of the can.

Diameter _____

Volume _____

Explanation

This question would be scored with the three-point scale. First, measure the line segment with your ruler. It's 4 inches long. So the radius of the can is 4 inches. Now, for parts A and B, you need to find the diameter and volume of the can. Be sure to show all work in the space provided, and explain everything you did in part B. Because the diameter is twice the radius, and the radius is 4 inches, the diameter is 8 inches. Make sure you fill that in right away, and write in something like, "I multiplied the radius by 2." Now, the formula for the volume of a cylinder is $V = \pi r^2 h$. (The formula is on the reference sheet.) Remember, $V = Bh$, and the base is a circle, which is πr^2. Fill in what you know.

$$V = \pi r^2 h$$
$$V = (3.14)(4)^2(13)$$
$$V = (3.14)(16)(13)$$
$$V = 653.12$$

The volume of the cylinder is 653.12 cubic inches. Your explanation would be something like, "I used the formula for the volume of a cylinder; filled in the values of the radius, π, and the height; and solved for V."

41 On the number line below, graph the following inequality:

$$9 < x \le 17$$

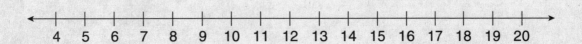

Explanation

This question would be scored with the two-point scale. The key to this question is remembering when you use the empty circle and when you use the solid one. You use the empty circle for < and >, and you use the solid circle for $\le$ and $\ge$. So the graph of the above inequality looks like this.

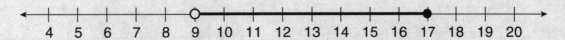

You could draw it right on the line, or above it, as long as you use the empty circle and the solid circle in the right places.

42 Grace and Chang have 36 cups of lemonade to sell. Grace sold $\frac{2}{3}$ of the lemonade, and Chang sold $\frac{1}{4}$ as many cups as Grace sold. How many cups of lemonade did Chang sell?

Show your work.

Answer _____

Explanation

This question would be scored with the two-point scale. Be sure to show all work in the space provided. And of course, explain in as much detail as you can. There is a total of 36 cups of lemonade to sell. Grace sells $\frac{2}{3}$ of that.

$$\frac{2}{3} \times \frac{\overset{12}{36}}{\underset{1}{1}} = 24$$

Chang sold $\frac{1}{4}$ as many as Grace.

$$\frac{2}{3}_{1} \times \frac{36^{12}}{1} = 24$$

The answer is 6 cups of lemonade.

43 The triangle below, *RST,* is a right triangle.

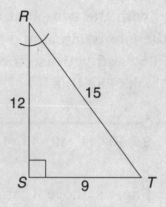

What are the numerical values of the sine and cosine of angle *R?*

Sine R _____

Cosine R _____

Explanation

This question would be scored with the two-point scale. Remember the magic word: SOHCAHTOA. The sine of *R* is equal to the opposite, which is 9, over the hypotenuse, which is 15. That's $\frac{9}{15}$, which reduces to $\frac{3}{5}$. You could also call it 0.6. The cosine of *R* is equal to the adjacent, which is 12, over the hypotenuse, which is 15. That's $\frac{12}{15}$, which reduces to $\frac{4}{5}$. You could also call it 0.8.

---

44 Fill in the missing numbers in the table below for the ordered pairs for the function $3x + y = 8$.

Part A

x	y
0	
	2
-2	
	-1

Part B

On the following grid, graph the function $3x + y = 8$, labeling the points with the coordinates from the table above:

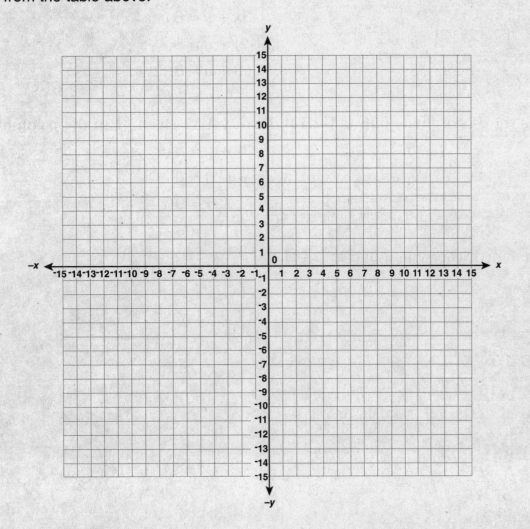

Explanation

This question would be scored with the three-point scale. Use the numbers in the table in the equation $3x + y = 8$. First, you're given 0 in the x column.

$$3x + y = 8$$

$$(3)(0) + y = 8$$
$$0 + y = 8$$
$$y = 8$$

So the first coordinate pair is (0,8). Next, you're given 2 in the y column.

$$3x + y = 8$$
$$3x + 2 = 8$$
$$3x = 6$$
$$x = 2$$

So the second coordinate pair is (2,2). Next, you're given –2 in the x column.

$$3x + y = 8$$
$$(3)(-2) + y = 8$$
$$-6 + y = 8$$
$$y = 14$$

So the third coordinate pair is (–2,14). Next, you're given –1 in the y column.

$$3x + y = 8$$
$$3x + (-1) = 8$$
$$3x = 9$$
$$x = 3$$

So the fourth coordinate pair is (3,–1).

For part B, you have to plot the points (0,8), (2,2), (−2,14), and (3,−1). It should look like this.

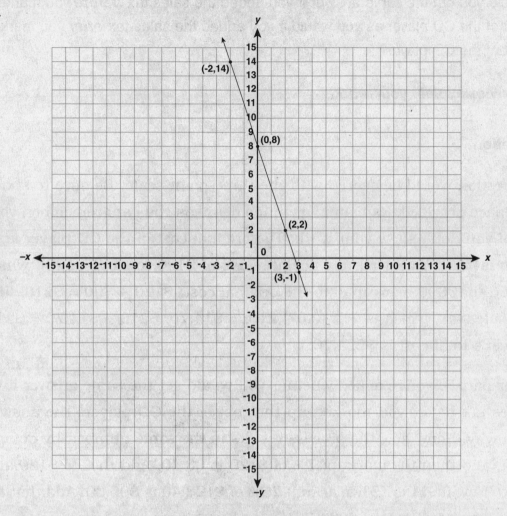

Label the points on the line in parentheses, connect them, and put some arrows on the ends to show that this line is infinite. The more information you include, the higher you'll score.

45 A CD player cost the electronics store $120. The store manager marks up the cost of the CD player by 75% to get the price. The sales tax on the CD player is 7%.

Part A

How much would it cost for you to buy the CD player, all totaled, including tax?

Show your work.

Answer _____

Would you get the same answer if you added the sales tax *before* you marked up the cost of the CD player as you would if you added the sales tax *after* you marked up the cost of the CD player?

Show or explain your work.

Explanation

This question would be scored with the three-point scale. Be sure to show all work in the space provided. Explain in as much detail as you can; remember, you can lose points if you don't show your work. The original cost of the CD player is $120. The markup is 75%. You need to calculate 75% of $120, or (0.75)(120), which is $90. Now, add that $90 markup onto the original cost: $120 + $90 = $210. Finally, add on the sales tax, which is 7%. (0.07)(210) = $14.70. $210 + $14.70 = $224.70. The CD player's total cost is $224.70.

Now, for part B, you're asked whether you would get the same answer if you added the sales tax *before* you marked up the cost of the CD player. The answer is YES. Explain why. Show how the process results in the same number, by doing the sales tax first, and then the markup: 7% of $120 is $8.40. Add that to $120 and you get $128.40. Now find the 75% markup: 75% of $128.40 is $96.30. Add that to $128.40 and you get a total cost of $224.70.

Now that you've gone over the first practice test, go back to the review section and review the topics of the questions you missed. If necessary, look up these topics in your textbook. Then take the second practice test.

PRACTICE TEST TWO

1   Which operation should be performed first in order to solve the following equation?

$$4 + 27 \div 3(10 - 2) =$$

A   addition
B   subtraction
C   multiplication
D   division

2   Which one of these represents the prime factorization of 54?

F $3 \times 2 \times 9$
G $2^2 \times 3^2$
H 2×3^3
J 6×3^2

3   Karl has a bag of jelly beans. He gives Jenna $\frac{1}{4}$ of his jelly beans, Mimi $\frac{1}{6}$ of his jelly beans, Alan $\frac{2}{5}$ of his jelly beans, and Paula $\frac{1}{5}$ of his jelly beans.

Which of these shows the correct order of Karl's friends, from the person with the fewest jelly beans to the person with the most?

A   Jenna, Alan, Paula, Mimi
B   Alan, Jenna, Paula, Mimi
C   Mimi, Paula, Jenna, Alan
D   Paula, Mimi, Jenna, Alan

4   To make the following equation true, what is the value of r?

$$r^5 = 243$$

F   2
G   3
H   4
J   5

5   Monica and Charleton did a puzzle on Saturday. Monica did $\frac{1}{5}$ of the puzzle, and later Charleton did 30% of the puzzle. What percentage of the puzzle did Monica and Charleton do all together?

A   40%
B   50%
C   60%
D   80%

Session 1, Part 1

Go to next page

6 A drawing of a parallelogram is shown below. *WX* measures 28 inches, and *WY* measures half the length of WX. What is the perimeter of parallelogram *WXYZ*?

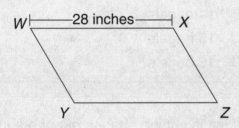

F 14 inches
G 42 inches
H 56 inches
J 84 inches

7 Brandon read $\frac{5}{8}$ of the newspaper. What percentage of the newspaper did Brandon read?

A 5.8%
B 37.5%
C 58%
D 62.5%

8 Use the ruler to solve this problem.

A corral shaped like a rectangle is shown below.

scale: 1 centimeter = 1 meter

Based on the scale, what is the perimeter of the corral?

F 8.5 meters
G 16 meters
H 17 meters
J 17.5 meters

9   In the diagram below, trace over the line segments *AB* and *EF*. Which angle is formed by the intersection of line segments *AB* and *EF*?

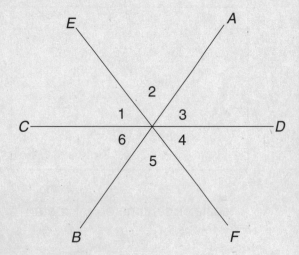

A   1
B   2
C   3
D   4

10 Cynthia and Mick are sealing envelopes for a mailing. Cynthia says, "I will have sealed 7 envelopes if I seal this one and two more." Mick says, "And I will have sealed 6 envelopes if I seal this one and one more." Which of these is a correct statement?

F   Cynthia has sealed 7 envelopes.
G   Mick has sealed 3 envelopes.
H   Cynthia has sealed 3 envelopes.
J   Mick has sealed 4 envelopes.

11 Courtney is buying bagels. The first bagel costs her $0.40, and each additional bagel costs her $0.32. If Courtney spends $4.88 on bagels, how many did she buy?

A   13 bagels
B   14 bagels
C   15 bagels
D   16 bagels

12 Liza is cutting shapes out of a square, as shown.

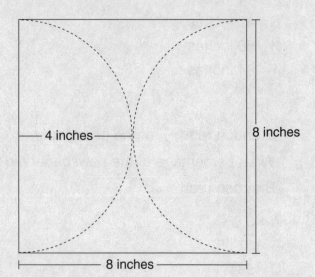

What is the perimeter of the shape Liza has left after she makes the cuts shown above?

(use π = 3.14)

F   28.56 inches
G   41.12 inches
H   44.56 inches
J   57.12 inches

Session 1, Part 1

Go to next page

The Ridgewood High School Baseball Team

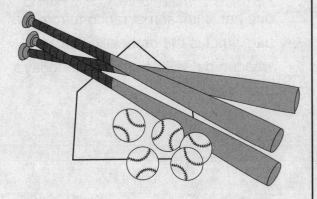

Directions: Numbers 13 through 16 are about the Ridgewood High School baseball team.

13 In order to get people to come watch the baseball team on Opening Day, Ridgewood High is offering a special deal on snacks, as shown in the table below.

Hot Dog/Chips Deal	
Buy this many hot dogs	**Get this many bags of chips free**
1	2
2	3
3	5
4	8
5	12

If Chris receives 17 free bags of chips with the hot dogs he buys for his friends on Opening Day, and the pattern continues, how many hot dogs did he buy?

A 6
B 7
C 8
D 9

14 The Ridgewood High baseball team practices after school Monday through Friday for 3 hours, and on Saturdays for 5 hours. How many hours does the team practice during a 16-week season that starts on a Sunday?

F 140 hours
G 160 hours
H 240 hours
J 320 hours

Session 1, Part 1

Go to next page

15 The coach of the Ridgewood High baseball team is keeping track of the number of home runs each of her players hits in the table below.

Home Runs	
Player	Number of Home Runs
Jake	5
Justine	2
Daryl	4
Anthony	6
Roger	0
Ramona	7
Billy	2
Christian	5
Mikey	5

What is the *mean* number of home runs hit by the players on the team?

A 4
B 4.5
C 5
D 7

16 The team has two sets of bats they use for practice. There are 15 wooden bats and 9 metal bats. All of these bats, and only these bats, are kept in a bag. If one bat is chosen at random from the bag, what is the probability that a wooden bat will be chosen for today's practice?

F $\dfrac{5}{8}$

G $\dfrac{3}{8}$

H $\dfrac{3}{5}$

J $\dfrac{5}{3}$

Session 1, Part 1

Go to next page

17 The figure below is two parallel lines cut by a transversal.

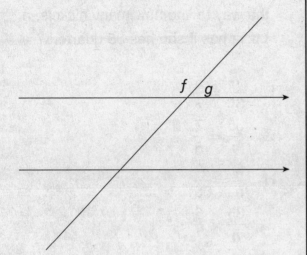

Which of these statements is true about angles *f* and *g*?

A Angles *f* and *g* are complementary.
B Angles *f* and *g* are supplementary.
C Angles *f* and *g* are vertical angles.
D Angles *f* and *g* are right angles.

18 Which value for *x* will make the statement below true?

$$2(x - 7) + 4 = 10$$

F −4
G 9
H 10
J 17

19 Which of these inequalities is represented on the number line below?

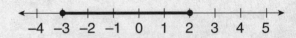

A $-3 < x < 2$
B $-3 \leq x \leq 2$
C $-3 \leq x < 2$
D $-3 < x \leq 2$

20 Jenna is keeping track of how much money she spends on baseball cards in one week.

Daily Baseball Card Purchases	
Day	**Money Spent**
Sunday	$4.50
Monday	$2.35
Tuesday	$3.98
Wednesday	$4.71
Thursday	$3.87
Friday	$4.03
Saturday	$3.57

What is the *median* amount of money Jenna spent on baseball cards this week?

F $3.57
G $3.86
H $3.98
J $4.03

Session 1, Part 1

Go to next page

21 A bowl in David's tool room is divided into equal compartments, and each compartment is filled with some hardware.

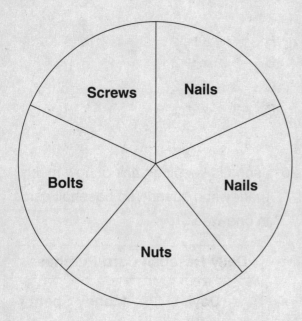

If David reaches into the bowl without looking, what is the probability that he will pull a bolt out of the bowl?

A $\dfrac{1}{16}$

B $\dfrac{1}{25}$

C $\dfrac{1}{5}$

D $\dfrac{1}{4}$

22 Lucia is counting her change, and she has 5 quarters for every 7 nickels she has. Which of these proportions shows the way to find how many nickels, n, Lucia has if she has 35 quarters?

F $\dfrac{n}{35} = \dfrac{5}{7}$

G $\dfrac{35}{n} = \dfrac{7}{5}$

H $\dfrac{5}{n} = \dfrac{7}{35}$

J $\dfrac{35}{n} = \dfrac{5}{7}$

23 Which of the following is a true statement?

A Two shapes that are **similar** are always exactly the same size.

B Two shapes that are **congruent** are never exactly the same size.

C Two shapes that are **similar** are not always exactly the same size.

D Two shapes that are **congruent** have only one side in common.

Session 1, Part 1

Go to next page

24 Heather buys a box of computer diskettes. She promised her brother Bruce five of the diskettes. If h is the number of diskettes in the unopened box Heather buys, and b is the number of diskettes Heather has after she gives Bruce five of them, which of the following equations can be used to find out how many diskettes Heather has after she gives Bruce five of them?

F $b = h - 5$
G $h = b - 5$
H $b - h = 5$
J $b = 5 + h$

25 Mo works at the laundromat after school, and he gets paid $4.00 per hour. He also receives $0.20 for each pound of laundry he washes. If he works for 4 hours on Thursday, and makes a total of $18.40, what is the average number of pounds of laundry he washes per hour on Thursday?

A 2
B 3
C 4
D 12

26 A butcher must put out the meat for the day. She knows the following: There are only pieces of beef, pork, lamb, and chicken. There are three times as many pieces of pork as beef, twice as many pieces of chicken as lamb, and half as many pieces of lamb as beef. If there are 5 pieces of beef, how many pieces of meat must the butcher put out?

F 27.5
G 40
H 50
J 50.5

Session 1, Part 1

Go to next page

27 On the grid below, one triangle is shown. Two coordinates for another triangle are also shown.

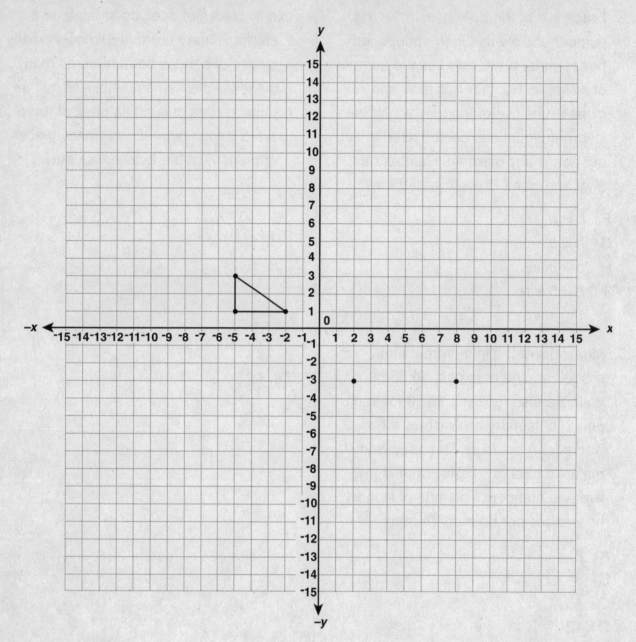

Which of these sets of coordinates will complete a second triangle that is similar to the triangle shown?

A   (8,–5)

B   (8,–6)

C   (8,–4)

D   (8,–7)

SESSION 1, PART 2

28 Which of the following numbers could replace the variable x in the following inequality:

$$0.81 > x > \frac{3}{5}$$

Circle all of the numbers that would make this inequality true.

$\frac{4}{8}$ $\quad$ 0.68 $\quad$ $\frac{5}{6}$ $\quad$ $\frac{3}{4}$ $\quad$ $\frac{5}{4}$

Explain why each number you circled could replace the variable x.

____

____

____

____

____

____

____

Session 1, Part 2

Go to next page

$$s \div v = w$$

$$w \times t = t$$

$$s + t = s$$

$$w + w = s$$

In the four equations above, variables *s, t, v,* and *w* each represent a whole number. If *w* = 2, find the value for each of the remaining variables. Show all of your work.

Show your work.

Answers

t = _____

s = _____

v = _____

Session 1, Part 2

Go to next page

30 Raul works at an appliance store during the summer. He is keeping track of the number of fans and air conditioners he sells during the summer months in the table below.

Summer Fan and A/C Sales		
Month	Number of Fans	Number of Air Conditioners
May	31	19
June	37	23
July	47	29
August	45	27

Part A

Construct a double line graph of the information from the table on the grid below, being sure to

- title the graph and label the axes
- be accurate
- make a key
- be consistent with the scale

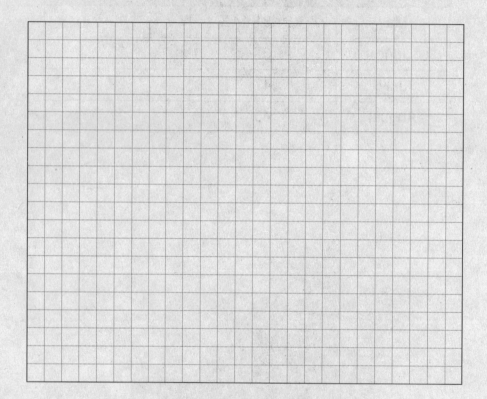

Part B

Using your double line graph, determine which month the difference between the number of fans sold and the number of air conditioners sold was the **smallest.**

Month _____

31 These are the coordinates for quadrilaterals X, Y, and Z.

Quadrilateral X: (−6,2), (−6,6), (−10,6), (−10,2)

Quadrilateral Y: (−3,−1), (−9,−1), (−3,−3), (−9,−3)

Quadrilateral Z: (3,2), (6,2), (3,−4), (6,−6)

Part A

On the following grid, plot, connect, and label the coordinates of quadrilaterals X, Y, and Z:

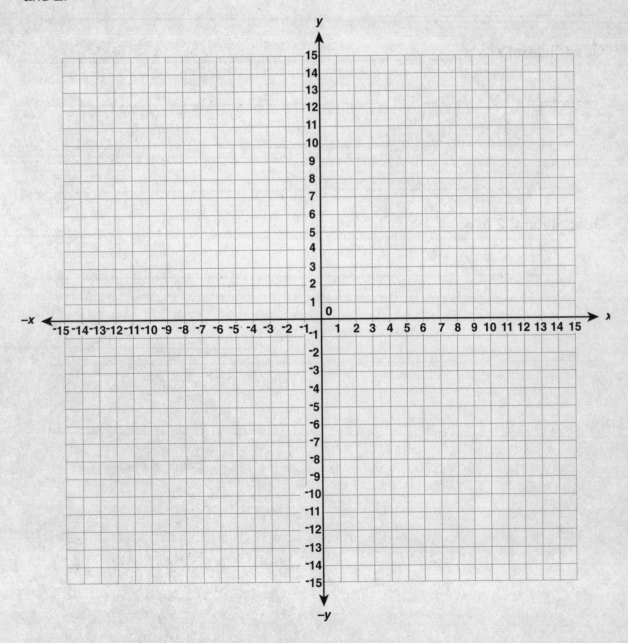

Part B

Name and describe what type(s) of quadrilateral each one is, and explain in words those qualities that helped you identify each.

Quadrilateral X is a ______________________________

____

____

____

Quadrilateral Y is a ______________________________

____

____

____

Quadrilateral Z is a ______________________________

____

____

____

Session 1, Part 2

Go to next page

32 Bernice is looking through some magazines to read on her upcoming train ride. There are three categories of magazines: sports, entertainment, and fashion.

Magazines		
Sports	**Entertainment**	**Fashion**
Baseball Daily	*Movie World*	*Attraction*
Soccer Today	*TV Time*	*Fad*
	Video Review	

How many different combinations of magazines, each consisting of one sports, one entertainment, and one fashion, can Berniece choose?

Show your work.

Answer ________________

Bernice is late for her train, so she chooses her three magazines at random. What is the probability that Bernice chooses one of the sports magazines, *MovieWorld,* and *Fad*?

Probability ________________

Session 1, Part 2

Go to next page

33 The tree in the picture is 9 feet high, and it casts a shadow that is 12 feet long. What is the distance from the end of the tree to the top of the shadow it casts?

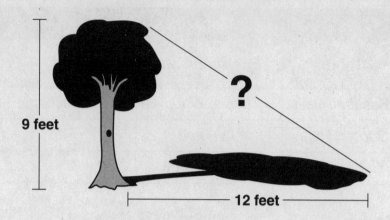

Show or explain your work.

Answer _____

34 Every year, Martin earns $36,000. On average, how much does Martin earn per month?

Show your work.

Answer _____

Go to next page

35 Jacques baked 5 batches of chocolate chip cookies and 4 batches of brownies for a bake sale. Each batch contains the same number of items. Jacques baked a total of 360 items.

Part A

Write an equation to figure out how many items are in each batch (*b*).

Equation

Part B

Use the equation you wrote to figure out how many items are in each batch.

Show your work.

Answer _____

Color Games at Camp Crestwood

This Sunday is the annual Color Games competition. There will be lots of activities and competitions, and the campers' parents have been invited to watch the action. The camp counselors are in charge of organizing the events.

Directions: Numbers 36 through 38 are all about Color Games at Camp Crestwood.

36 Camp Crestwood's owners give the counselors $700 to have T-shirts made up for Sunday. They are to use 65% of the money on the actual shirts, and the rest on the logos for the shirts. If each logo costs $1.75, how many T-shirts can the counselors buy?

Show your work.

Answer _____

Session 2

Go to next page

37 One Color Games activity was a kickball competition. The campers were divided into 5 squads: the Red Squad, the Blue Squad, the Green Squad, the White Squad, and the Purple Squad. Each squad had to play every other squad one time. How many kickball games were played?

Show your work.

Answer ________________

Session 2

Go to next page

38 Another activity was a swimming competition in the lake, the path of which is marked off by flags. The swimmers start at coordinates (2,2). Swimmers make their next turn at the flag at (7,2). Then they turn at the flag at (7,7), and then again at the flag at (12,7). The end of the swimming competition is at (12,2).

Part A

On the grid below, show the path the swimmers take to get from the starting point to the end. Plot the points of the swimming path as described above, connect them, and label each point.

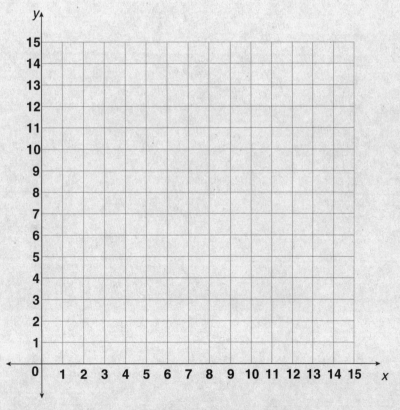

Session 2

Go to next page

Part B

The distance from the starting point to the first turn is 90 feet. How long is the entire swimming path?

Answer _____

Explain how you got your answer.

_____

_____

_____

_____

_____

_____

39 Hannah makes handmade jewelry as a hobby, and she decides to try selling her pieces. If she doesn't sell 1,500 pieces by the end of two years, she will give up on her jewelry business. She's keeping track of how many pieces she sells per month on the following table:

Jewelry Sales	
Month	**Number of Pieces Sold**
January	71
February	79
March	74
April	77

If Hannah keeps selling her jewelry at about the same rate, ESTIMATE how many *more* months it will take until she reaches her goal.

Show your work.

Answer ______________

Session 2

Go to next page

40 Use the ruler to help answer this question.

A cylinder-shaped bucket has a height twice the length of the line segment below.

_____

Part A

If π is equal to 3.14, and the volume of the bucket is 471 cubic inches, what are the radius and the diameter of the bucket in inches?

Show your work.

Diameter _____

Volume _____

Part B

Explain how you calculated the radius and the diameter of the bucket.

Radius _____

_____

Diameter _____

_____

41   On the number line below, graph the following inequality:

$$2 < x \le 12$$

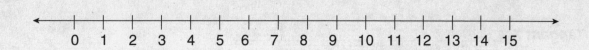

42   Clint and Tanya have 27 widgets to put together at a factory. Clint has put together $\frac{4}{9}$ of the widgets, and Tanya has put together $\frac{2}{3}$ as many widgets as Clint has. How many widgets has Tanya put together?

Show your work.

Answer _____

43 The triangle below, *XYZ*, is a right triangle.

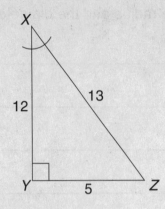

What are the numerical values of the sine and tangent of angle *X*?

Sine X _____

Tangent X _____

44 Fill in the missing numbers in the table below for the ordered pairs for the function
$3x + y = 9$.

Part A

x	y
0	
	0
2	
	-3

Part B

On the following grid, graph the function $3x + y = 9$, labeling the points with the coordinates from the table above:

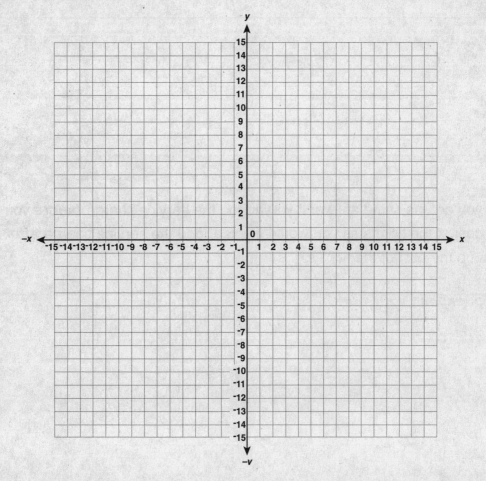

45 Tickets for a concert cost the ticket agency $40 each. The agent marks up the cost of each ticket by 80% before they are made available to consumers. A service charge is added on to all tickets purchased by consumers. The service charge on the tickets is 10%.

Part A

How much would it cost a consumer to buy a ticket, all totaled, including the service charge?

Show your work.

*Answer*_____

Part B

Would you get the same answer if you added the service charge *before* you marked up the cost of the tickets as you would if you added the service charge *after* you marked up the cost of the tickets?

Show or explain your work.

Session 2

Stop

PRACTICE TEST TWO ANSWERS AND EXPLANATIONS

ANSWER KEY FOR MULTIPLE-CHOICE QUESTIONS

1.	B	10.	J	19.	B
2.	H	11.	C	20.	H
3.	C	12.	G	21.	C
4.	G	13.	B	22.	J
5.	B	14.	J	23.	C
6.	J	15.	B	24.	F
7.	D	16.	F	25.	B
8.	H	17.	B	26.	F
9.	B	18.	H	27.	D

SESSION I, PART I

1   Which operation should be performed first in order to solve the following equation?

$$4 + 27 \div 3(10 - 2) =$$

A   addition
B   subtraction
C   multiplication
D   division

B According to the order of operations (PEMDAS), whatever's inside parentheses must be taken care of first. Because the operation inside the parentheses is subtraction, the correct answer choice is B. Notice that you don't have to solve the equation to get the answer.

2   Which one of these represents the prime factorization of 54?

F $3 \times 2 \times 9$
G $2^2 \times 3^2$
H 2×3^3
J 6×3^2

H This is a good opportunity to use the Process of Elimination. Prime factorization means that all of the numbers must be prime. So eliminate any answer choice that contains numbers that are not prime. That gets rid of F, because 9 is not prime, and J, because 6 is not prime. Now you're left with two choices, G and H. Check out G. 2^2 is 4, and 3^2 is 9. $4 \times 9 = 36$. That's not 54. The correct answer choice must be H, because $3^3 = 27$ and $27 \times 2 = 54$.

3 Karl has a bag of jelly beans. He gives Jenna $\frac{1}{4}$ of his jelly beans, Mimi $\frac{1}{6}$ of his jelly beans, Alan $\frac{2}{5}$ of his jelly beans, and Paula $\frac{1}{5}$ of his jelly beans. Which of these shows the correct order of Karl's friends, from the person with the fewest jelly beans to the person with the most?

A Jenna, Alan, Paula, Mimi
B Alan, Jenna, Paula, Mimi
C Mimi, Paula, Jenna, Alan
D Paula, Mimi, Jenna, Alan

C Basically, this question is asking you to order the fractions $\frac{1}{4}$, $\frac{1}{6}$, $\frac{2}{5}$, and $\frac{1}{5}$. One way to do this is to compare the fractions in pairs. But you might also realize that if you're comparing a few fractions that all have the same numerator (in this case, 1), the bigger the denominator, the smaller the fraction. That means that $\frac{1}{6}$ is the smallest. Next comes $\frac{1}{5}$, then $\frac{1}{4}$. Finally, $\frac{2}{5}$ is the biggest. Now just match those fractions to the appropriate person from the questions (in other words, Jenna = $\frac{1}{4}$, Mimi = $\frac{1}{6}$, etc.).

4 To make the following equation true, what is the value of r?

$$r^5 = 243$$

F 2
G 3
H 4
J 5

G This is a great opportunity to use the answer choices to your advantage. Start with choice F by making $r = 2$. Does $2^5 = 243$? $2 \times 2 \times 2 \times 2 \times 2 = 32$. $32 \neq 243$. Now try G by making $r = 3$. Does $3^5 = 243$? $3 \times 3 \times 3 \times 3 \times 3 = 243$. That's it.

5   Monica and Charleton did a puzzle on Saturday. Monica did $\frac{1}{5}$ of the puzzle, and later Charleton did 30% of the puzzle. What percentage of the puzzle did Monica and Charleton do all together?

A   40%

B   50%

C   60%

D   80%

B Because the question asks for a percentage, convert the fraction $\frac{1}{5}$ to a percentage.

$$\frac{1}{5} = \frac{x}{100}$$

$$5x = 100$$

$$\frac{5x}{5} = \frac{100}{5}$$

$$x = 20\%$$

So the total percentage is 20% + 30%, which is 50%.

6   A drawing of a parallelogram is shown below. *WX* measures 28 inches, and *WY* measures half the length of *WX*. What is the perimeter of parallelogram *WXYZ*?

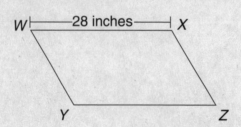

F   14 inches

G   42 inches

H   56 inches

J   84 inches

J Perimeter is the sum of the lengths of all the sides. The question provides you with the length of the parallelogram, 28 inches, and *WY*, which is half of 28, or 14 inches. In a parallelogram, the opposite sides are equal, so *YZ* must be 28 and *XZ* must be 14. So, 28 + 14 + 28 + 14 = 84 inches.

7   Brandon read $\frac{5}{8}$ of the newspaper. What percentage of the newspaper did Brandon read?

   A   5.8%

   B   37.5%

   C   58%

   D   62.5%

   D Because the answer asks for a percentage, all you have to do is convert $\frac{5}{8}$ to a percentage: $\frac{5}{8} = \frac{x}{100}$. Cross multiply and you get $8x = 500$. Divide both sides by 8 to isolate x, and you will get $x = 62.5\%$.

8   Use the ruler to solve this problem.

   A corral shaped like a rectangle is shown below.

   scale: 1 centimeter = 1 meter

   Based on the scale, what is the perimeter of the corral?

   F   8.5 meters

   G   16 meters

   H   17 meters

   J   17.5 meters

   H Use the ruler to measure the sides of the picture. You should get a length of 5 centimeters and a width of 3.5 centimeters. Perimeter is the sum of the lengths of the figure—in this case 5 + 3.5 + 5 + 3.5, which is 17. Because the scale tells that 1 centimeter equals 1 meter, the answer is 17 meters.

9   In the diagram below, trace over the line segments *AB* and *EF*. Which angle is formed by the intersection of line segments *AB* and *EF*?

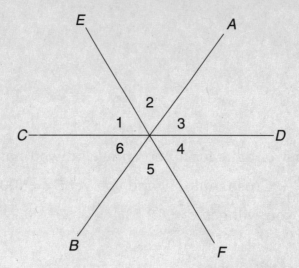

A   1

B   2

C   3

D   4

B The angles formed at the intersection of line segments *AB* and *EF* are angles 2 and 5. Because 5 is not among the answer choices, the answer must be 2.

10   Cynthia and Mick are sealing envelopes for a mailing. Cynthia says, "I will have sealed 7 envelopes if I seal this one and two more." Mick says, "And I will have sealed 6 envelopes if I seal this one and one more." Which of these is a correct statement?

F   Cynthia has sealed 7 envelopes.

G   Mick has sealed 3 envelopes.

H   Cynthia has sealed 3 envelopes.

J   Mick has sealed 4 envelopes.

J Start with Cynthia. She says she'll have sealed 7 envelopes if she seals the one she's holding and two more. The one she's holding, plus the two more, is three. So if she adds three to what she already has sealed, she must already have sealed 4. Stop and check the answer choices: What do they say about Cynthia? If they don't say she has sealed 4 envelopes, you can eliminate them. That gets rid of F and H. Now figure out Mick's situation. He says he'll have sealed 6 envelopes if he seals the one he's holding and one more. The one he's holding, plus the one more, is two. So if he adds two to what he already has sealed, he must already have sealed 4. That's J.

11 Courtney is buying bagels. The first bagel costs her $0.40, and each additional bagel costs her $0.32. If Courtney spends $4.88 on bagels, how many did she buy?

A   13 bagels

B   14 bagels

C   15 bagels

D   16 bagels

C The bagels cost Courtney a total of $4.88. Subtract away that $0.40 for the first bagel right away (but don't forget about it, because it counts as a bagel!). $4.88 – $0.40 = $4.48. Now, take that $4.48 and divide it by the cost of each additional bagel, $0.32, to get the number of remaining bagels. $4.48 ÷ $0.32 = 14 bagels. But wait, don't pick B. You first have to add on that one bagel from that "$0.40 for the first bagel." So the answer is 15 bagels.

12 Liza is cutting shapes out of a square, as shown.

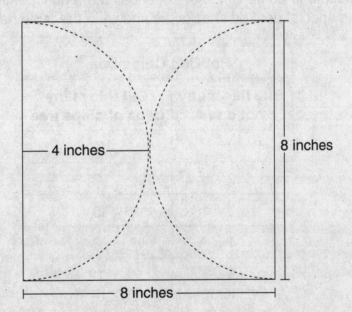

What is the perimeter of the shape Liza has left after she makes the cuts shown above?

(use π = 3.14)

F   28.56 inches

G   41.12 inches

H   44.56 inches

J   57.12 inches

G Liza is starting out with a square piece of paper, with sides of 8 inches. The two sides that aren't affected by the cutting are still 8, and there are two of them, so that's 16 inches. Now, Liza's cutting out two semicircles, each with a radius of 4. Those two semicircles would actually form one circle if they were each flipped upside down. The formula for the circumference of a circle is $C = 2\pi r$. The radius is 4, so solve for C.

$$C = 2\pi r$$

$$C = (2)(3.14)(4)$$

$$C = 25.12 \text{ inches}$$

Now add up what you have: 16 inches + 25.12 inches = 41.12 inches.

The Ridgewood High School Baseball Team

Directions: Numbers 13 through 16 are about the Ridgewood High School baseball team.

13 In order to get people to come watch the baseball team on Opening Day, Ridgewood High is offering a special deal on snacks, as shown in the table below.

Hot Dog/Chips Deal	
Buy this many hot dogs	**Get this many bags of chips free**
1	2
2	3
3	5
4	8
5	12

If Chris receives 17 free bags of chips with the hot dogs he buys for his friends on Opening Day, and the pattern continues, how many hot dogs did he buy?

B Look for the pattern in the free bags of chips. The difference between 2 and 3 is 1. The difference between 3 and 5 is 2. The difference between 5 and 8 is 3. The difference 8 and 12 is 4. So the pattern increases by 1 every additional hot dog. That means that the next number of free bags of chips (which would correspond with 6 hot dogs) would be 12 + 5, or 17, and the next (which would correspond with 7 hot dogs) would be 17 + 6, or 23. That's the right answer.

14 The Ridgewood High baseball team practices after school Monday through Friday for 3 hours, and on Saturdays for 5 hours. How many hours does the team practice during a 16-week season that starts on a Sunday?

F 140 hours
G 160 hours
H 240 hours
J 320 hours

J Each week, the baseball team practices 5 days for 3 hours and 1 day for 5 hours. That's 15 + 5, or 20 hours a week. 20 hours × 16 weeks is 320 hours.

15 The coach of the Ridgewood High baseball team is keeping track of the number of home runs each of her players hits in the table below.

Home Runs	
Player	Number of Home Runs
Jake	5
Justine	2
Daryl	4
Anthony	6
Roger	0
Ramona	7
Billy	2
Christian	5
Mikey	5

What is the *mean* number of home runs hit by the players on the team?

A 4
B 4.5
C 5
D 7

A To find the mean, which is the same as the average, add up the total, and divide by the number of things. $5 + 2 + 4 + 6 + 0 + 7 + 2 + 5 + 5 = 36$. Now, divide 36 by the number of hitters, which is 9 (don't forget to include Roger, even though he hit 0 home runs). $36 \div 9 = 4$. The median and mode of the home runs hit by the players are both 5.

16 The team has two sets of bats they use for practice. There are 15 wooden bats and 9 metal bats. All of these bats, and only these bats, are kept in a bag. If one bat is chosen at random from the bag, what is the probability that a wooden bat will be chosen for today's practice?

F $\dfrac{5}{8}$

G $\dfrac{3}{8}$

H $\dfrac{3}{5}$

J $\dfrac{5}{3}$

F There are a total of 24 bats in the bag. There are 15 wooden bats. So, the probability that a wooden bat will be chosen from the bag is $\dfrac{15}{24}$. Divide both parts of the fraction by the GCF, 3, to simplify $\dfrac{15}{24}$ to $\dfrac{5}{8}$.

17 The figure below is two parallel lines cut by a transversal.

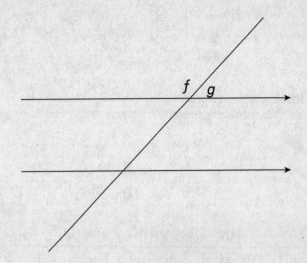

Which of these statements is true about angles *f* and *g*?

A   Angles *f* and *g* are complementary.
B   Angles *f* and *g* are supplementary.
C   Angles *f* and *g* are vertical angles.
D   Angles *f* and *g* are right angles.

B Angles *f* and *g* lie on the same line, so that means the measures of their angles add up to 180 degrees. That means they are supplementary. Complimentary angles are angles whose measures add up to 90 degrees.

18 Which value for *x* will make the statement below true?

$$2(x - 7) + 4 = 10$$

F   −4
G   9
H   10
J   17

H Your job is to solve for *x*. First, subtract 4 from both sides.

$$2(x - 7) + 4 = 10$$
$$\underline{\ -4\quad -4}$$
$$2(x - 7)\quad = 6$$

Then divide both sides by 2.

$$\frac{2(x-7)}{2} = \frac{6}{2}$$
$$x - 7 = 3$$

Then add 7 to both sides.

$$
\begin{array}{r}
x - 7 = 3 \\
+7 \quad +7 \\
\hline
x = 10
\end{array}
$$

You also could have used the Distributive Property on the original equation, $2(x - 7) + 4 = 10$. Distribute that 2 to both the numbers in the parentheses by multiplying $2 \times x$ and 2×-7. The equation becomes $2x - 14 + 4 = 10$. That can be simplified into $2x - 10 = 10$. Then solve for x.

$$
\begin{array}{r}
2x - 10 = 10 \\
+10 \quad +10 \\
\hline
2x = 20 \\
\frac{2x}{2} = \frac{20}{2} \\
x = 10
\end{array}
$$

19 Which of these inequalities is represented on the number line below?

$$-4 \ -3 \ -2 \ -1 \ 0 \ 1 \ 2 \ 3 \ 4 \ 5$$

A $-3 < x < 2$

B $-3 \le x \le 2$

C $-3 \le x < 2$

D $-3 < x \le 2$

B The solid circles on the number line represent either less than or equal to ($\le$) or greater than or equal to ($\ge$). The bold line connects -3 to 2. So the number line represents all the numbers between and including -3 and 2. That's B. But just by noticing that both circles are solid on the number line, you could have eliminated answer choices A, C, and D because they all contain either < or > (which are represented by empty circles).

20 Jenna is keeping track of how much money she spends on baseball cards in one week.

Daily Baseball Card Purchases	
Day	Money Spent
Sunday	$4.50
Monday	$2.35
Tuesday	$3.98
Wednesday	$4.71
Thursday	$3.87
Friday	$4.03
Saturday	$3.57

What is the *median* amount of money Jenna spent on baseball cards this week?

F $3.57
G $3.86
H $3.98
J $4.03

H The median is the middle number in a set when the numbers are listed in numerical order. To find the median in this set of numbers, start by putting the numbers in order from least to greatest: $2.35, $3.57, $3.87, $3.98, $4.03, $4.50, $4.71. Now, which number is directly in the middle? It's $3.98.

21 A bowl in David's tool room is divided into equal compartments, and each compartment is filled with some hardware.

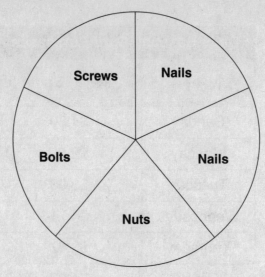

If David reaches into the bowl without looking, what is the probability that he will pull a bolt out of the bowl?

A $\dfrac{1}{16}$

B $\dfrac{1}{25}$

C $\dfrac{1}{5}$

D $\dfrac{1}{4}$

C Don't let the fact that there are two compartments full of nails fool you; there are five compartments in the bowl, and bolts are in one of them. Also, the question says that the compartments are of equal size. So, the probability that David will pick a bolt out of the bowl is 1 out of 5, or $\dfrac{1}{5}$.

22 Lucia is counting her change, and she has 5 quarters for every 7 nickels she has. Which of these proportions shows the way to find how many nickels, n, Lucia has if she has 35 quarters?

F $\dfrac{n}{35} = \dfrac{5}{7}$

G $\dfrac{35}{n} = \dfrac{7}{5}$

H $\dfrac{5}{n} = \dfrac{7}{35}$

J $\dfrac{35}{n} = \dfrac{5}{7}$

J The key to setting up a proportion is being careful to put like terms in the same places.

$$\frac{5 \text{ quarters}}{7 \text{ nickels}} = \frac{35 \text{ quarters}}{n \text{ nickels}}$$

That proportion is the same thing as the one in answer choice J. Notice that you're not asked to actually solve for n. Why do more work than you have to?

23 Which of the following is a true statement?

    A   Two shapes that are ***similar*** are always exactly the same size.
    B   Two shapes that are ***congruent*** are never exactly the same size.
    C   Two shapes that are ***similar*** are not always exactly the same size.
    D   Two shapes that are ***congruent*** have only one side in common.

C In this case, you must go through the answers and evaluate them. Notice that they are either about the definition of **similar** or the definition of **congruent.** Similar figures have the same shape, but may differ in size. That eliminates A. Congruent means having the same shape and size. That gets rid of B and D. That leaves us with C. Similar triangles can differ in size.

24 Heather buys a box of computer diskettes. She promised her brother Bruce five of the diskettes. If h is the number of diskettes in the unopened box Heather buys, and b is the number of diskettes Heather has after she gives Bruce five of them, which of the following equations can be used to find out how many diskettes Heather has after she gives Bruce five of them?

    F $b = h - 5$
    G $h = b - 5$
    H $b - h = 5$
    J $b = 5 + h$

F The question states that h is the number of diskettes in the box to begin with. Heather gives five of the diskettes to Bruce. Finally, b represents the number of diskettes in the box *after* the five diskettes are given to Bruce. So the number of diskettes in the box, minus 5, equals the number after the five are given away. In other words, $h - 5 = b$. That's the same as choice F.

25 Mo works at the laundromat after school, and he gets paid $4.00 per hour. He also receives $0.20 for each pound of laundry he washes. If he works for 4 hours on Thursday, and makes a total of $18.40, what is the average number of pounds of laundry he washes per hour on Thursday?

A 2

B 3

C 4

D 12

B Mo made a total of $18.40. The question states that he gets $4.00 an hour, and he worked 4 hours that day. So $16 (that's 4 hours × $4.00) out of that $18.40 represents his hourly wage. That leaves $2.40, which was gotten from the $0.20 per pound of laundry. First divide $2.40 by $0.20 to get 12. So Mo did 12 pounds of laundry on Thursday. But the question is asking for the average number of pounds of laundry he washed per hour. Again, he worked 4 hours, and because average is equal to total divided by number, just divide the total pounds of laundry, 12, by the number of hours, or 4. That gives you 3.

26 A butcher must put out the meat for the day. She knows the following: There are only pieces of beef, pork, lamb, and chicken. There are three times as many pieces of pork as beef, twice as many pieces of chicken as lamb, and half as many pieces of lamb as beef. If there are 5 pieces of beef, how many pieces of meat must the butcher put out?

F 27.5

G 40

H 50

J 50.5

F Turn those sentences into equations: There are three times as many pieces of pork as beef, so $P = 3B$. There are twice as many pieces of chicken as lamb, so $C = 2L$. There are half as many pieces of lamb as beef, so $L = \dfrac{B}{2}$. There are 5 pieces of beef, so $B = 5$. Now start solving by substituting what you already know. Because B is 5, substitute the 5 into the equation $L = \dfrac{B}{2}$ to solve for L.

$$L = \frac{B}{2}$$

$$L = \frac{5}{2}$$

$$L = 2.5$$

So there are $2\frac{1}{2}$ pieces of lamb. Now you can substitute 2.5 for L in the equation $C = 2L$ to solve for C.

$$C = 2L$$

$$C = 2(2.5)$$

$$C = 5$$

So there are 5 pieces of chicken. Now substitute 5 for B in the equation $P = 3B$ to find P.

$$P = 3B$$

$$P = 3(5)$$

$$P = 15$$

So there are 15 pieces of pork.

Now add up the meat: 5 (beef) + 2.5 (lamb) + 5 (chicken) + 15 (pork) = 27.5 total pieces of meat.

27 On the grid below, one triangle is shown. Two coordinates for another triangle are also shown.

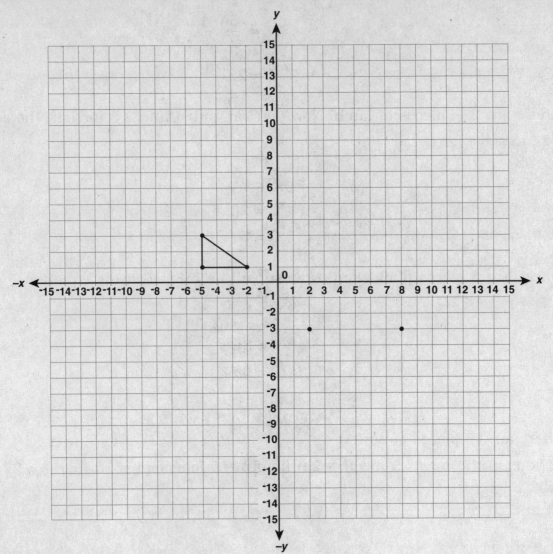

Which of these sets of coordinates will complete a second triangle that is similar to the triangle shown?

A (8,–5)

B (8,–6)

C (8,–4)

D (8,–7)

D First, count the boxes to get the line lengths of the shown triangle: It has a base of 3 and a height of 2. Now connect the dots on that other line. Count the boxes to find its length: 6. That line forms a part of a triangle that has to be similar to the first triangle. The 6 has a relationship to the 3: It's twice as much. So the height of the second triangle should be twice 2, the height of the first triangle. That would be 4. Now look at the answer choices. They all have an x-coordinate of 8. Since the missing coordinate should form a line that is 4 boxes long, count 4 boxes up or down from (8,–3). You wind up at either (8,–1) or (8,–7). Though (8,–1) isn't among the choices, (8,–7) is—it's D.

It is also possible that the two given coordinates represent the vertical line from the first triangles. (That would mean that the new triangle is rotated at 90 degrees.) If that were the case, the new triangle would be three times the size because the line measuring 6 would be similar to the line measuring 2. Therefore, the other line, 3, would be 9 in the new triangle. The third set of coordinates would then have to be (8,6) or (8,–12). Neither of those coordinates is listed in the answer choices.

28 Which of the following numbers could replace the variable x in the following inequality:

$$0.81 > x > \frac{3}{5}$$

Circle all of the numbers that would make this inequality true.

$\frac{4}{8}$ 0.68 $\frac{5}{6}$ $\frac{3}{4}$ $\frac{5}{4}$

Explain why each number you circled could replace the variable x.

Explanation

This question would be scored with the two-point scale. For the first part of the question, you need to circle all the numbers that could replace x in the inequality $0.81 > x > \frac{3}{5}$. One way to start is to convert $\frac{3}{5}$ to 0.6 using your calculator. Now you know you're looking for anything between 0.81 and 0.6. $\frac{4}{8}$ is the same as 0.5, which is too small. 0.68 works, so that one gets circled. $\frac{5}{6}$ is the same as $0.83\overline{3}$, which is too big. $\frac{3}{4}$ is the same as 0.75, which works, so that one gets circled. $\frac{5}{4}$ is greater than 1, so that doesn't work. So 0.68 and $\frac{3}{4}$ should be circled. If you find fractions easier to work with than decimals, you could have converted everything to fractions and compared them that way. Either method is correct.

For the second part of the question, you have to explain how you got your answers. Explain in as much detail as you can. You can either explain it in sentences or mathematically, such as "I circled 0.68 and $\frac{3}{4}$ because they are both between 0.81 and $\frac{3}{5}$, which is the same as 0.6," and "$\frac{4}{8} < \frac{3}{5}$, so this doesn't work," and so forth. Your explanation would have to include something about how the circled numbers are between 0.81 and $\frac{3}{5}$.

$$s \div v = w$$

$$w \times t = t$$

$$s + t = s$$

$$w + w = s$$

In the four equations above, variables *s, t, v,* and *w* each represent a whole number. If $w = 2$, find the value for each of the remaining variables. Show all of your work.

Show your work.

Answers

t = _____

s = _____

v = _____

Explanation

This question would be scored with the two-point scale. Be sure to show your work in the space provided. Explain in as much detail as you can. Because you're told that $w = 2$, look for an equation that involves a w. How about $w + w = s$? Make that $2 + 2 = s$, which means that $s = 4$. Fill that in where it says "Answers" so you don't forget. If $s = 4$, look at $s + t = s$. That's $4 + t = 4$. So, t must equal 0. Fill that in where it says "Answers" so you don't forget. Knowing that $t = 0$ makes sense with the equation $w \times t = t$ as well. Now, there's another equation with s: $s \div v = w$. In other words, $4 \div v = 2$. That means that v must equal 2. Fill that in where it says "Answers" so you don't forget. So $s = 4$, $t = 0$, and $v = 2$.

30 Raul works at an appliance store during the summer. He is keeping track of the number of fans and air conditioners he sells during the summer months in the table below.

Summer Fan and A/C Sales		
Month	Number of Fans	Number of Air Conditioners
May	31	19
June	37	23
July	47	29
August	45	27

Part A

Construct a double line graph of the information from the table on the grid below, being sure to

- title the graph and label the axes
- be accurate
- make a key
- be consistent with the scale

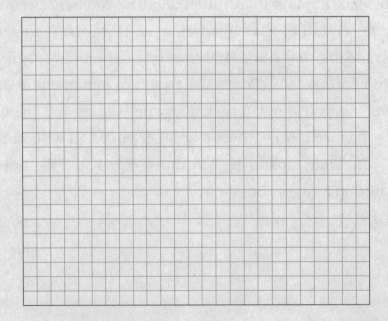

Using your double line graph, determine which month the difference between the number of fans sold and the number of air conditioners sold was the ***smallest.***

Month _____

Explanation

This question would be scored with the thee-point scale. In part A, your job is to construct a graph based on the information you're given about the fan and air conditioner sales. Here's what it should look like.

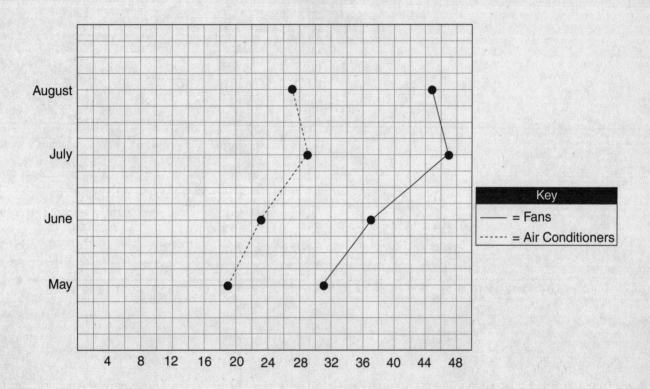

The *x*- and *y*-axes could be reversed in your drawing and it would still be correct. Besides the actual information about the fans and air conditioners, your graph must also include an appropriate title, correct labels for the axes, a consistent scale, and an appropriate key.

For part B, your job is to determine the month during which the difference between the number of fans sold and the number of air conditioners sold was the ***smallest.*** That month would be May; those "May" dots are the closest together.

31 These are the coordinates for quadrilaterals X, Y, and Z.

Quadrilateral X: (–6,2), (–6,6), (–10,6), (–10,2)
Quadrilateral Y: (–3,–1), (–9,–1), (–3,–3), (–9,–3)
Quadrilateral Z: (3,2), (6,2), (3,–4), (6,–6)

Part A

On the following grid, plot, connect, and label the coordinates of quadrilaterals X, Y, and Z:

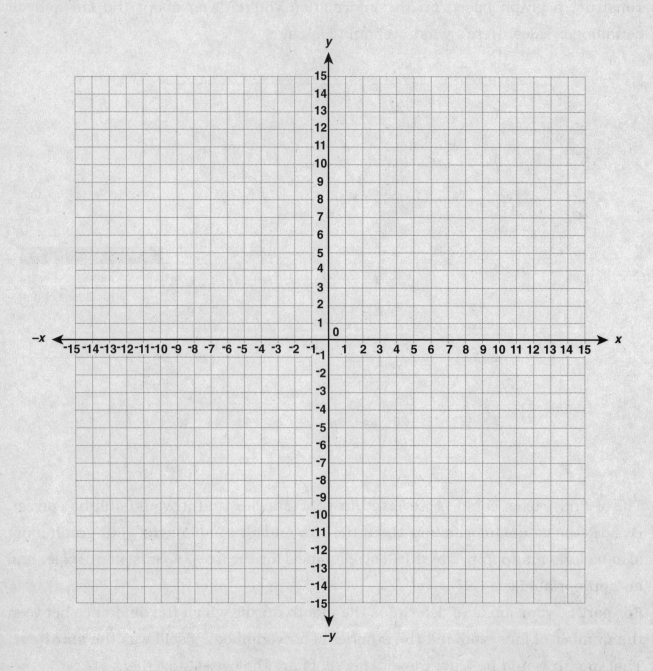

Roadmap to 8th Grade Math: New York Edition

Name and describe what type(s) of quadrilateral each one is, and explain in words those qualities that helped you identify each.

Quadrilateral X is a _____

Quadrilateral Y is a _____

Quadrilateral Z is a _____

Explanation

This question would be scored with the 3-point scale. For part A, your job is to graph these quadrilaterals. That should look like this (note that each point is labeled).

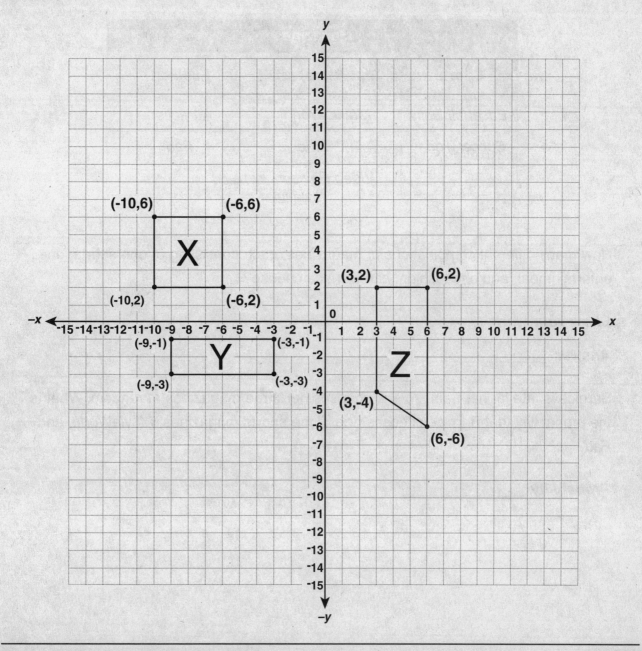

For part B, you need to identify each quadrilateral and explain how you did so. Explain in as much detail as you can. Quadrilateral X is a square because each pair of opposite sides is parallel, all sides are congruent, and there are four right angles. You could also point out that quadrilateral X is also considered a rectangle and a parallelogram. Quadrilateral Y is a rectangle because it has two pairs of opposite sides that are parallel and four right angles. You could also point out that quadrilateral Y is considered a parallelogram. Quadrilateral Z is a trapezoid because it has one pair of opposite sides that are parallel.

32 Bernice is looking through some magazines to read on her upcoming train ride. There are three categories of magazines: sports, entertainment, and fashion.

Magazines		
Sports	**Entertainment**	**Fashion**
Baseball Daily	Movie World	Attraction
Soccer Today	TV Time	Fad
	Video Review	

How many different combinations of magazines, each consisting of one sports, one entertainment, and one fashion, can Bernice choose?

Show your work.

Answer _____

Bernice is late for her train, so she chooses her three magazines at random. What is the probability that Bernice chooses one of the sports magazines, *MovieWorld*, and *Fad*?

Probability _____

Explanation

This question would be scored with the two-point scale. Be sure to show all work in the space provided. Explain in as much detail as you can; remember, when in doubt, write it out. You could make a tree diagram like the one below.

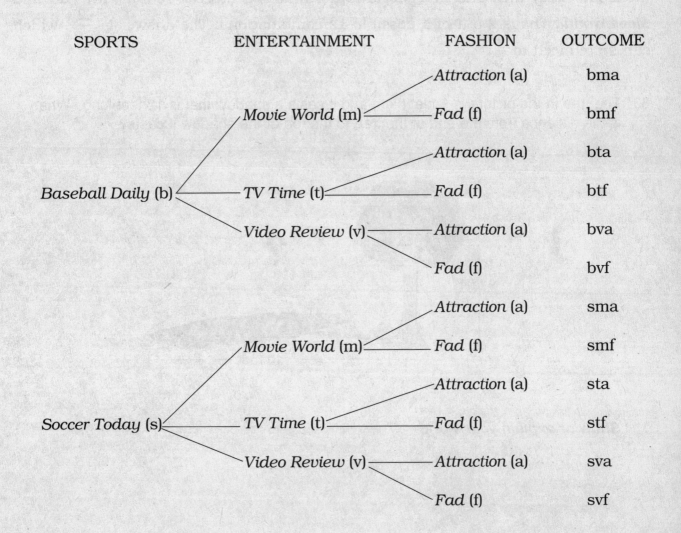

| SPORTS | ENTERTAINMENT | FASHION | OUTCOME |

There are 12 different combinations, or outcomes. You also could have used the Counting Principle: If event M can occur in m ways and is followed by event N that can occur in n ways, then the event M followed by the event N can occur in $m \times n$ ways. In the case of the magazines, that's the number of choices for sports magazines, which is 2, times the number of choices for entertainment magazines, which is 3, times the number of choices for fashion magazines, which is 2. So, $2 \times 3 \times 2 = 12$.

For the second part of the question, you need to figure out the probability that Bernice randomly chooses one of the sports magazines, *MovieWorld*, and *Fad*. The first part of the question showed that there are 12 possible combinations. The question is, how many of them include both *MovieWorld* and *Fad*? There's bmf (*Baseball Daily* with *Fad* and *MovieWorld*), and smf (*Soccer Today* with *Fad* and *MovieWorld*). That's 2 out of a possible 12 combinations. The answer is $\frac{2}{12}$, which can be reduced to $\frac{1}{6}$.

33 The tree in the picture is 9 feet high, and it casts a shadow that is 12 feet long. What is the distance from the end of the tree to the top of the shadow it casts?

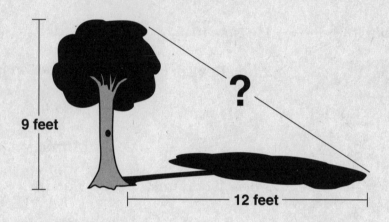

Show or explain your work.

Answer _____

Explanation

This question would be scored with the two-point scale. Be sure to use the space provided to show or explain your work. As always, explain in as much detail as you can. You're really looking for the hypotenuse of a right triangle formed by the height of the tree, 9 feet, and the length of the shadow, 12 feet. Use the Pythagorean theorem to find the length of the missing side, which is also the number of feet in the hypotenuse.

$$c^2 = a^2 + b^2$$
$$c^2 = 9^2 + 12^2$$
$$c^2 = 81 + 144$$
$$c^2 = 225$$
$$c = \sqrt{225}$$
$$c = 15$$

The answer is 15 feet (don't forget the word "feet"). You could also have solved this by recognizing that in a right triangle with a legs of 9 and 12, the hypotenuse would have to be 15 because the 9-12-15 right triangle is a multiple of the primitive Pythagorean triple, 3-4-5. That would also be a valid explanation.

SESSION 2

34 Every year, Martin earns $36,000. On average, how much does Martin earn per month?

Show your work.

Answer _____

Explanation

This question would be scored with the two-point scale. Be sure to show all work in the space provided. Explain in as much detail as you can. This is a proportion question. Martin earns $36,000 every year. Because the question asks how much he earns per month, set up the proportion as dollars per month. In other words, instead of $36,000 per 1 year, make it $36,000 per 12 months:

$$\frac{\$36,000}{12 \text{ months}} = \frac{x}{1 \text{ month}}$$

Now cross multiply, and you get $12x = \$36,000$. Divide both sides by 12, and you get $3,000. The answer is $3,000 per month (don't forget to add "per month").

35 Jacques baked 5 batches of chocolate chip cookies and 4 batches of brownies for a bake sale. Each batch contains the same number of items. Jacques baked a total of 360 items.

Part A

Write an equation to figure out how many items are in each batch (b).

Equation _____

Part B

Use the equation you wrote to figure out how many items are in each batch.

Show your work.

Answer _____

Explanation

This question would be scored with the two-point scale. For part A, you have to write an equation to figure out how many items are in each batch. Jacques baked 5 batches of cookies and 4 batches of brownies, for a total of 360 items. So, $5b + 4b = 360$, because b is the same amount whether it's cookies or brownies. In other words, $9b = 360$.

For part B, you get to solve for b. Be sure to show all work in the space provided. Explain in as much detail as you can.

$$9b = 360$$

$$\frac{9b}{9} = \frac{360}{9}$$

$$b = 40$$

There are 40 items in a batch.

Color Games at Camp Crestwood

This Sunday is the annual Color Games competition. There will be lots of activities and competitions, and the campers' parents have been invited to watch the action. The camp counselors are in charge of organizing the events.

Directions: Numbers 36 through 38 are all about Color Games at Camp Crestwood.

36 Camp Crestwood's owners give the counselors $700 to have T-shirts made up for Sunday. They are to use 65% of the money on the actual shirts, and the rest on the logos for the shirts. If each logo costs $1.75, how many T-shirts can the counselors buy?

Show your work.

Answer _____

Explanation

This question would be scored with the two-point scale. Be sure to show all work in the space provided and explain in as much detail as you can. The counselors have $700, and they are to use 65% of the money on the shirts and the rest on logos. Stop there. If they use 65% for the shirts, what percent do they use for logos? 35%. Now, what is 35% of $700? Remember, you can use your calculator on this section (but write down what you've calculated!), so just punch in 0.35 × $700, and you get $245. So the counselors spend $245 on the logos. The logos cost $1.75 each, so to find out how many shirts the counselors can buy, divide $245 by $1.75. You will get 140. The answer is 140 shirts.

37 One Color Games activity was a kickball competition. The campers were divided into 5 squads: the Red Squad, the Blue Squad, the Green Squad, the White Squad, and the Purple Squad. Each squad had to play every other squad one time. How many kickball games were played?

Show your work.

Answer _____

Explanation

This question would be scored with the two-point scale. Be sure to show all work in the space provided. Explain in as much detail as you can. You could make a tree diagram.

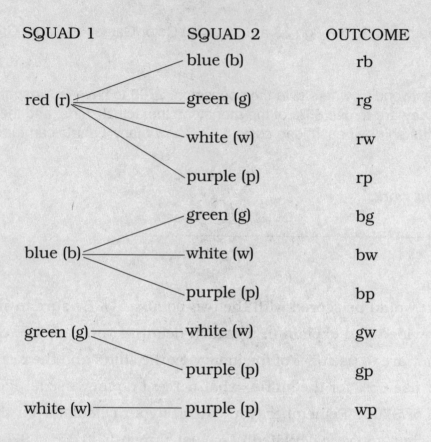

The answer is 10.

This is a combination, so you can also use the formula $_nC_r = \dfrac{n!}{(n-r)!\,(r!)}$. There are 5 squads playing 2 at a time. Plug that into the formula.

$$_5C_2 = \dfrac{5!}{(5-2)!\,(2!)}$$

$$\dfrac{5 \times 4 \times 3 \times 2 \times 1}{(3 \times 2 \times 1)(2 \times 1)}$$

Don't forget to reduce.

$$\dfrac{5 \times 4}{2 \times 1} = \dfrac{20}{2} = 10$$

The answer is 10.

38 Another activity was a swimming competition in the lake, the path of which is marked off by flags. The swimmers start at coordinates (2,2). Swimmers make their next turn at the flag at (7,2). Then they turn at the flag at (7,7), and then again at the flag at (12,7). The end of the swimming competition is at (12,2).

Part A

On the grid below, show the path the swimmers take to get from the starting point to the end. Plot the points of the swimming path as described above, connect them, and label each point.

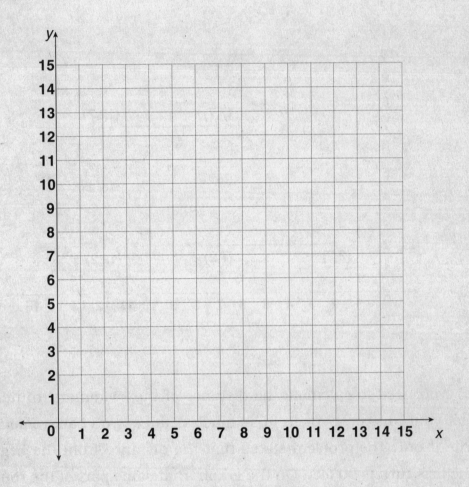

Part B

The distance from the starting point to the first turn is 90 feet. How long is the entire swimming path?

Answer _____

Explain how you got your answer.

Explanation

This question would be scored with the three-point scale. For part A, your job is to graph the coordinates of each turn described in the directions. This is what it should look like (note that each point is labeled).

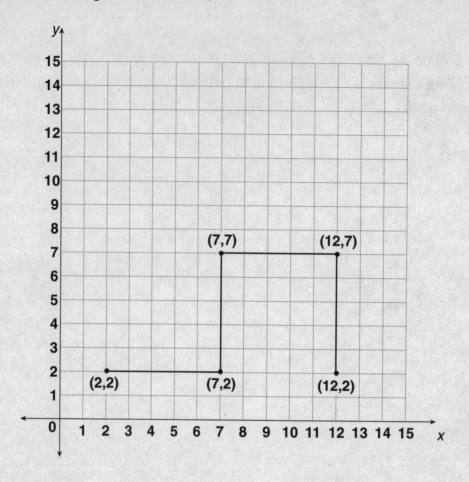

For part B, you need to calculate the distance of the swimmers' route, and explain how you got your answer. Explain in as much detail as you can; remember, when in doubt, write it out. The problem states that the distance from the beginning of the route to the first turn is 90 feet. On the graph, that same part of the route is 5 boxes. So the scale is 5 boxes = 90 feet. That means that each box equals 18 feet. Now count how many boxes make up the entire route from start to finish. It's 20 boxes. If 1 box = 18 feet, 20 boxes = 360 feet. The answer is 360 feet.

You also could have figured that since the scale is 5 boxes = 90 feet and there are 20 boxes, the proportion is $\frac{5}{90} = \frac{20}{x}$. Cross multiply and you get $5x = 1,800$. The answer is still 360 feet.

39 Hannah makes handmade jewelry as a hobby, and she decides to try selling her pieces. If she doesn't sell 1,500 pieces by the end of two years, she will give up on her jewelry business. She's keeping track of how many pieces she sells per month on the following table:

Jewelry Sales	
Month	**Number of Pieces Sold**
January	71
February	79
March	74
April	77

If Hannah keeps selling her jewelry at about the same rate, ESTIMATE how many *more* months it will take until she reaches her goal.

Show your work.

Answer _____

Explanation

This question would be scored with the two-point scale. Be sure to show all work in the space given. As always, explain in as much detail as you can. Note that you're being asked to ESTIMATE. You could start by rounding to an approximate monthly sales amount, say, 75. Hannah has 1,500 pieces to sell, so $1,500 \div 75$ is 20 months. She's already sold jewelry for 4 months, so that leaves 16 months.

There's another way to solve this problem: You could start by rounding to an approximate monthly sales amount, say, 75 pieces. She sold jewelry for four months, so that's about 75×4, or 300. Hannah needs to sell 1,200 more pieces. If it takes her 4 months to sell 300 pieces, how many months would it take to sell the remaining 1,200? 300×4 is 1,200, so $4 \times 4 = 16$. The answer is still 16 months.

40 Use the ruler to help answer this question.

A cylinder-shaped bucket has a height twice the length of the line segment below.

_____

Part A

If π is equal to 3.14, and the volume of the bucket is 471 cubic inches, what are the radius and the diameter of the bucket in inches?

Show your work.

Diameter _____

Volume _____

Part B

Explain how you calculated the radius and the diameter of the bucket.

Radius _____

Diameter _____

Explanation

This question would be scored with the three-point scale. First, measure the line segment with your ruler. It's 3 inches long. The height of the bucket is twice that, or 6 inches. Now, for parts A and B, you need to find the radius and diameter of the bucket. Be sure to show all work in the space provided, and explain everything you did in part B. Explain in as much detail as you can; remember, when in doubt, write it out. The formula for the volume of a cylinder is $V = \pi r^2 h$. Remember, $V = Bh$, and the base is a circle, which is πr^2. Fill in the information that you know.

$$V = \pi r^2 h$$

$$471 = (3.14)(r^2)(6)$$

$$471 = (18.84)(r^2)$$

$$\frac{471}{18.84} = \frac{18.84r^2}{18.84}$$

$$25 = r^2$$

$$\sqrt{25} = \sqrt{r^2}$$

$$5 = r$$

The radius of the cylinder is 5 inches. Your explanation would be something like, "I used the formula for the volume of a cylinder; filled in the values of the volume, π, and the height; and solved for r." Now, the diameter is twice the radius, so it must be 10 inches. Your explanation would be something like, "I got the diameter by doubling the radius."

41 On the number line below, graph the following inequality:

$$2 < x \leq 12$$

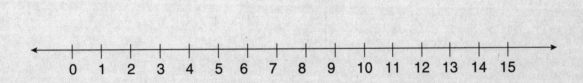

Explanation

This question would be scored with the two-point scale. The key to this question is remembering when you use the empty circle and when you use the solid one. You use the empty circle for < and >, and you use the solid circle for ≤ and ≥ . So the graph of the above inequality looks like this.

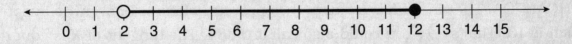

You could draw it right on the line, or above it, as long as you use the empty circle and the solid circle in the right places.

42 Clint and Tanya have 27 widgets to put together at a factory. Clint has put together $\frac{4}{9}$ of the widgets, and Tanya has put together $\frac{2}{3}$ as many widgets as Clint has. How many widgets has Tanya put together?

Show your work.

Answer _____

Explanation

This question would be scored with the two-point scale. Be sure to show all work in the space provided. It is important to explain in as much detail as you can. There is a total of 27 widgets to be put together. Clint put together $\frac{4}{9}$ of that. $\frac{4}{9}$ of 27 = 12. Tanya put together $\frac{2}{3}$ as many as Clint, which is $\frac{2}{3}$ of 12, or 8. The answer is 8 widgets.

43 The triangle below, *XYZ,* is a right triangle.

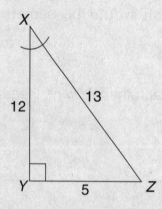

What are the numerical values of the sine and tangent of angle *X?*

Sine X _____

Tangent X _____

Explanation

This question would be scored with the two-point scale. Remember the magic word: SOHCAHTOA. The sine of *X* is equal to the opposite, which is 5, over the hypotenuse, which is 13. That's $\frac{5}{13}$. You could also call it 0.384. The tangent of *X* is equal to the opposite, which is 5, over the adjacent, which is 12. That's $\frac{5}{12}$. You could also call it 0.416.

44 Fill in the missing numbers in the table below for the ordered pairs for the function $3x + y = 9$.

Part A

x	y
0	
	0
2	
	-3

Part B

On the following grid, graph the function $3x + y = 9$, labeling the points with the coordinates from the table above:

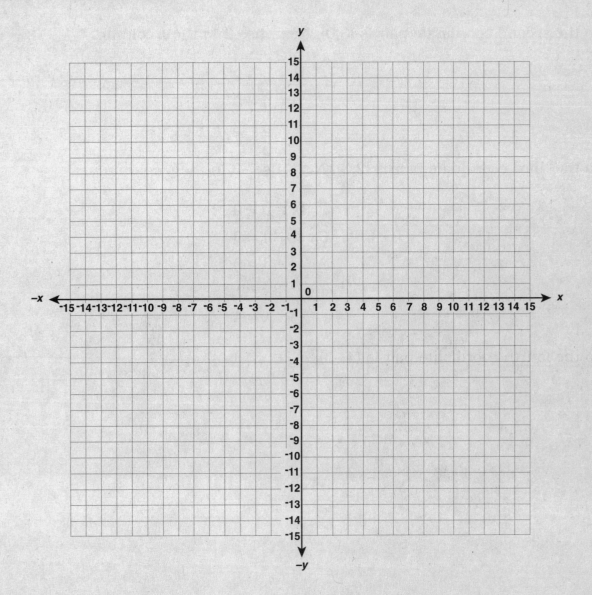

Explanation

This question would be scored with the three-point scale. Use the numbers in the table in the equation $3x + y = 9$. First, use 0 in the x column.

$$3x + y = 9$$
$$0 + y = 9$$
$$y = 9$$

So the first coordinate pair is (0,9). Next, use 0 in the y column.

$$3x + y = 9$$
$$3x + 0 = 9$$
$$3x = 9$$
$$x = 3$$

So the second coordinate pair is (3,0). Next, use 2 in the x column.

$$3x + y = 9$$
$$6 + y = 9$$
$$y = 3$$

So the third coordinate pair is (2,3). Next, use –3 in the y column.

$$3x + y = 9$$
$$3x + (-3) = 9$$
$$3x - 3 = 9$$
$$3x = 12$$
$$x = 4$$

So the fourth coordinate pair is (4,–3).

For part B, you have to plot the points (0,9), (3,0), (2,3), and (4,−3). It should look like this.

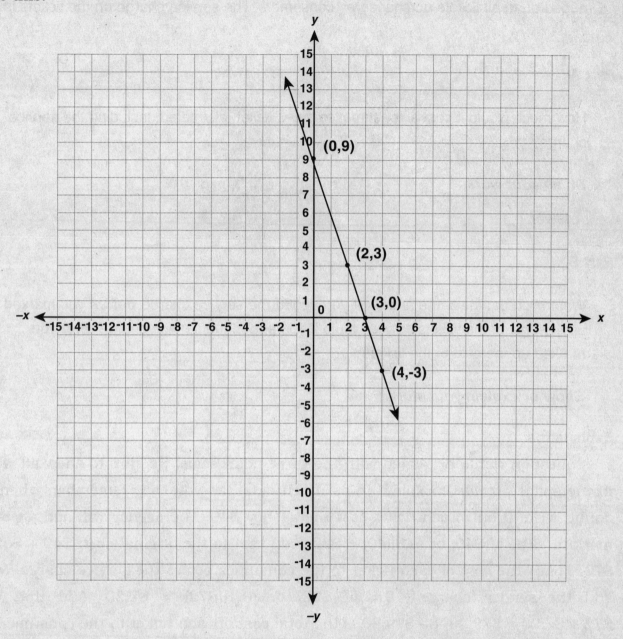

Label the points on the line in parentheses, connect them, and put some arrows on the ends to show that this line is infinite. The more information you include, the higher you'll score.

45 Tickets for a concert cost the ticket agency $40 each. The agent marks up the cost of each ticket by 80% before they are made available to consumers. A service charge is added on to all tickets purchased by consumers. The service charge on the tickets is 10%.

Part A

How much would it cost a consumer to buy a ticket, all totaled, including the service charge?

Show your work.

Answer _____

Part B

Would you get the same answer if you added the service charge **before** you marked up the cost of the tickets as you would if you added the service charge **after** you marked up the cost of the tickets?

Show or explain your work.

Explanation

This question would be scored with the three-point scale. Be sure to show all work in the space provided. Explain in as much detail as you can; remember, when in doubt, write it out. Each ticket cost the agency $40. The agency will add an 80% markup. 80% of $40, or (0.8)(40) = $32. Add that to the original cost: $40 + $32 = $72. That's the cost of each ticket to the consumer before the service charge. Now add the service charge: 10% of $72, or (0.1)($72) = $7.20. Add that on: $72 + $7.20 = $79.20. So $79.20 is the total cost of each ticket to the consumer.

Now, for part B, the question asks whether you'd get the same answer if you added the service charge *before* you marked up the cost of the ticket. The answer is YES. Explain why this is the case. Show how the process results in the same number, by doing the service charge first, and then the markup: 10% of $40 is $4. Add that onto $40 and you get $44. Now find the 80% markup: 80% of $44 is $35.20. Add that to $44 and you get a total cost of $79.20. The final cost is the same.

Find the Right School

BEST 351 COLLEGES
2004 EDITION
0-375-76337-6 • $21.95

COMPLETE BOOK OF COLLEGES
2004 EDITION
0-375-76330-9 • $24.95

COMPLETE BOOK OF
DISTANCE LEARNING SCHOOLS
0-375-76204-3 • $21.00

AMERICA'S ELITE COLLEGES
The Smart Buyer's Guide to the Ivy
League and Other Top Schools
0-375-76206-X • $15.95

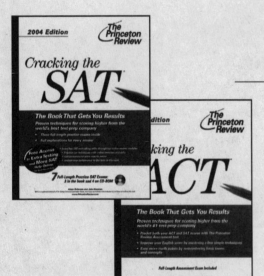

Get in

CRACKING THE SAT
2004 EDITION
0-375-76331-7 • $19.00

CRACKING THE SAT
WITH SAMPLE TESTS ON CD-ROM
2004 EDITION
0-375-76330-9 • $30.95

MATH WORKOUT FOR THE SAT
2ND EDITION
0-375-76177-2 • $14.95

VERBAL WORKOUT FOR THE SAT
2ND EDITION
0-375-76176-4 • $14.95

CRACKING THE ACT
2003 EDITION
0-375-76317-1 • $19.00

CRACKING THE ACT WITH
SAMPLE TESTS ON CD-ROM
2003 EDITION
0-375-76318-X • $29.95

CRASH COURSE FOR THE ACT
2ND EDITION
The Last-Minute Guide to Scoring High
0-375-75364-3 • $9.95

CRASH COURSE FOR THE SAT
2ND EDITION
The Last-Minute Guide to Scoring High
0-375-75361-9 • $9.95

Get Help Paying for it

DOLLARS & SENSE FOR COLLEGE STUDENTS
How Not to Run Out of Money by Midterms
0-375-75206-4 • $10.95

PAYING FOR COLLEGE WITHOUT GOING BROKE
2004 EDITION
0-375-76350-3 • $20.00

THE SCHOLARSHIP ADVISOR
5TH EDITION
0-375-76210-8 • $26.00

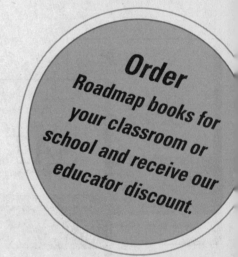